LĪLĀVATĪ OF BHĀSKARĀCĀRYA

LĪLĀVATĪ OF BHĀSKARĀCĀRYA

A Treatise of Mathematics of Vedic Tradition

with rationale in terms of modern mathematics largely based on N. H. Phadke's Marāthī translation of Līlāvatī

Translated by

KRISHNAJI SHANKARA PATWARDHAN
SOMASHEKHARA AMRITA NAIMPALLY
SHYAM LAL SINGH

MOTILAL BANARSIDASS PUBLISHERS
PRIVATE LIMITED • DELHI

Reprint: Delhi, 2006, 2008
First Edition: Delhi, 2001

ISBN: 978-81-208-1420-2 (Cloth)
ISBN: 978-81-208-1777-7 (Paper)

MOTILAL BANARSIDASS
41 U.A. Bungalow Road, Jawahar Nagar, Delhi 110 007
8 Mahalaxmi Chamber, 22 Bhulabhai Desai Road, Mumbai 400 026
236, 9th Main III Block, Jayanagar, Bangalore 560 011
203 Royapettah High Road, Mylapore, Chennai 600 004
Sanas Plaza, 1302 Baji Rao Road, Pune 411 002
8 Camac Street, Kolkata 700 017
Ashok Rajpath, Patna 800 004
Chowk, Varanasi 221 001

Printed in India
BY JAINENDRA PRAKASH JAIN AT SHRI JAINENDRA PRESS,
A-45 NARAINA, PHASE-I, NEW DELHI 110 028
AND PUBLISHED BY NARENDRA PRAKASH JAIN FOR
MOTILAL BANARSIDASS PUBLISHERS PRIVATE LIMITED,
BUNGALOW ROAD, DELHI 110 007

Foreword

The names 'Bhāskarācārya' and 'Līlāvatī' are well-known. If a student displays outstanding talent in Mathematics, the school teacher lovingly calls the student "Second Bhāskarācārya". Many legends about Līlāvatī are in vogue. It is but natural that Indians have a special interest in Bhāskarācārya and his works, especially the *Līlāvatī*. Bhāskarācārya was born in Maharashtra. Many translations of the *Līlāvatī* as well as commentaries on it are available. However, there is no work in Marāṭhī which studies this subject thoroughly from all angles in a modern — twentieth century — fashion. Professor N.H. Phadke has taken great pains in the preparation of the present work *A New Light on Līlāvatī*, thus filling a void.

Eighteen years ago Professor Phadke wrote a booklet on Indian Mathematicians. Therein we find a lot of authentic and interesting information about Bhāskarācārya and Līlāvatī. Doubtless the readers of that booklet will look forward expectantly at the present treatise.

In the *Līlāvatī*, Arithmetic is presented as an enjoyable playful activity. Professor Phadke has admirably succeeded in preserving this spirit in the *A New Light on Līlāvatī*. Since Professor Phadke studied many books in Sanskrit, Hindī, English and Marāṭhī, he could have easily given scholarly notes and references all over. But the author has resisted that temptation and has given the relevant supplementary information in six appendices. So the book is easy to read and the readers enjoy the playful atmosphere.

The *Līlāvatī* is a book on Arithmetic written in the twelfth century. It was used in India as a textbook for many centuries. Even now it is being used in Sanskrit Schools in some States. Techniques for the solution of problems are simple and easy to use and, moreover, there is a lot of interesting information in the problems presented therein.

Here, I would like to narrate my own experience. At the beginning of the *Sāndhyavandanam* (daily prayers), the Maharashtrians repeat twenty-four names of Lord Viṣṇu but in the version repeated in North India, there are some variations and less than twenty-four names. When I was a Professor at Benares Hindu University, I asked several scholars about the mystery of twenty-four names. None could give a satisfactory explanation. Finally the late Professor Vasudeva Sharan Agrawala of the College of Indology unravelled the mystery. Professor Phadke has given the same explanation in connection with Stanza CCLXX. In the four hands of Lord Viṣṇu there are a conch, a disc, a mace and a lotus. They can be placed in $4 \times 3 \times 2 \times 1 = 24$ ways in the four hands, thus giving rise to 24 different forms with 24 names. Similarly ten weapons can be placed in Lord Śiva's ten hands in $10! = 36,28,800$ ways. But one does not find so many names of Lord Śiva in the ancient literature.

Reader's attention is specially drawn to various forms of expressions used to address the student in different stanzas. Thus we find: "O, you intelligent girl *Līlāvatī* ", "O, friend", "My beloved", "Deer-eyed", "Fickle-eyed" etc. Scholars may draw varying conclusions from these regarding the types of students studying in the "Fun with Arithmetic" classes of Bhāskarācārya. Yet, one thing is crystal clear to the readers of the *Līlāvatī,* that learning begins with fun and flourishes in wonder. At the end of stanza LXXVIII we find, "if you know Arithmetic well, tell the number". When such a challenge appears at the ends of an interesting problem, the readers feel that it would have been wonderful to have been the members of Bhāskarācārya's class! In stanza CXXII, the following problem is given: "Suppose relishes are made by mixing 1, 2, 3, 4, 5 or 6 at a time from six substances which are respectively sweet, bitter, astringent, sour, salty and hot. O, arithmetician, tell me how many different relishes can be prepared?" One does not know whether boys and girls prepared such relishes but the readers relish, time and again, such "gourmet" Arithmetic!

Līlāvatī is such an ancient Arithmetic! Today we know many more techniques and results. In this regard, Professor Phadke has been quite alert. While explaining stanza CLXXV, he discusses the two cases (i) $cos(t) = 0$ and (ii) $cos(t) \neq 0$. This is an example of a result which was not known in the times of Bhāskarācārya but well-known now even to conscientious high school students. There is nothing surprising about this. Undergraduate students of Mathematics are now required to study several results which Newton did not know.

For those who are interested in both Sanskrit and Mathematics, *Līlāvatī* is an attractive illustrated arithmetized book of stories. But it would not be fair to merely consider *A New Light on Līlāvatī* as a cultural or literary treatise. In the history of Mathematics, *Līlāvatī* occupies an honoured place. It shows the extraordinary prowess of Bhāskarācārya who was not only a top Mathematician but also an excellent teacher. Plato says in *Republic:* "Arithmetic has a very great and elevating effect, compelling the soul to reason about abstract numbers, and if visible and tangible objects are intruding upon the argument, refusing to be satisfied."

If there be any truth in the above statement, then without getting distracted by interesting problems, poetic fancy and attractive descriptions, a child will be influenced by *Līlāvatī* and hopefully a new Bhāskarācārya will emerge.

In the history of Mathematics one does find such unusual incidents. The well-known Mathematician Madame Kowalewski was inspired by pages of an old Mathematics book which were used as wall paper! I'll not be surprised if some "Līlāvatī", inspired by the present treatise *A New Light on Līlāvatī,* will blossom forth into a top-notch researcher in Modern Mathematics.

Pune University
23-5-1971

V.V. Narlikar
Lokmanya Tilak
Professor of Applied Mathematics

Contents

Roman Transliteration of Devanāgarī

VOWELS

Short :	अ	इ	उ	ऋ	ऌ	(and ळ)	
	a	i	u	ṛ	ḷ		
Long :	आ	ई	ऊ	ए	ओ	ऐ	औ
	ā	ī	ū	e	o	ai	au

Anusvāra : ं = ṁ
Visarga : : = ḥ
Non-aspirant : ’ = ऽ

CONSONANTS

Classified :	क्	ख्	ग्	घ्	ङ्
	k	kh	g	gh	ṅ
	च्	छ्	ज्	झ्	ञ्
	c	ch	j	jh	ñ
	ट्	ठ्	ड्	ढ्	ण्
	ṭ	ṭh	ḍ	ḍh	ṇ
	त्	थ्	द्	ध्	न्
	t	th	d	dh	n

	प्	फ्	ब्	भ्	म्			
	p	**ph**	**b**	**bh**	**m**			
Un-Classified :	य्	र्	ल्	व्	श्	ष्	स्	ह्
	y	**r**	**l**	**v**	**ś**	**ṣ**	**s**	**h**
Compound :	क्ष्	त्र्	ज्ञ्					
	kṣ	**tr**	**jñ**					

Translators' Preface

It gives me great pleasure in placing before you this book, largely a translation of the *Līlāvatī Punardarśana* written in Marāṭhī by my revered Guru, the late Professor N.H. Phadke in 1971. He took great pains in writing this Maraṭhī translation with comments and explanations.

I must thank Smt. Subhadrabai Phadke for directing me to her eldest son Shri G.N. Phadke, Chief Engineer, Metropolitan Railway, Calcutta for the authority to publish this English translation, made by me, of Professor Phadke's *Līlāvatī Punardarśana.* I acknowledge the kindness shown by Mr. G.N. Phadke in giving me the authority.

It was some time in June 1982 that my student Somashekhar Naimpally expressed to me his desire to publish the English translation of the *Līlāvatī.* By the grace of the Almighty, I have been able to complete this translation before I completed 71 years of my life. I thank Somashekhar. I also wish to thank my wife Laxmibai *(alias* Mai) without whose cooperation I would not have been able to complete this work so soon.

Pune
November 2, 1982

K.S. Patwardhan

In 1977, I was inspired to translate *Līlāvatī* into English. Since I found the task difficult, I requested my esteemed teacher Professor Patwardhan, who not only readily agreed but finished the translation in record time. I went through his version, compared it with the original and made many changes. I have taken the liberty to shorten or simplify some derivations/proofs and to omit those that are found in current textbooks. It gives me a sense of fulfilment to have been able to play this role with my teachers Professors Patwardhan and Phadke who inspired me in my student days. It is a pleasure to express my gratitude.

Thunder Bay S.A. Naimpally
June 17, 1983

The *Līlāvatī* has been my first interest for the last two decades. I am happy that Professor Som A. Naimpally chose me to join as a translator for this book, and I owe him for the same.

It has been my endeavour to narrow down the gap between Sanskrit verses and their English renderings. Although it has been my ample endeavour that the English renderings should not be far away from the literary beauty of the verses, I have preferred to remain deeply faithful to the mathematical message. However, it seems that "faithfulness" and "beautifulness" rarely go together. As a consequence, renderings of certain stanze give only mathematical formulae. I have taken the liberty of making various alterations in the original typescript.

On behalf of the authors, I thank Professor K.V. Sarma of Madras for his high appreciation and suggestions to improve upon the original typescript. I record my appreciation and thank Mr. N.P. Jain and Dr. G.P. Bhatt for their interest and personal attention in bringing out this book and sharing some of the editing travails.

I believe that *Līlāvatī* can give pleasure and insight to every class of readers, both from children to senior citizens.

Rishikesh S.L. Singh
24-11-1991

Bhāskarācārya: His Life and Work

Rarely does one come across a person, at least in Maharashtra, who has not heard about Bhāskarācārya or Bhāskara II[1] — the great poet and mathematician. His book *Siddhāntaśiromaṇi* — especially the first part known as the *Līlāvatī* (Slate Mathematics or Arithmetic) — is well-known all over the world. Because of the poetic name the *Līlāvatī* and the excellent problems contained therein, Bhāskarācārya has earned the respect of scholars for the last eight hundred years.

The great nineteenth-century German mathematician Weierstrass said,

> "A mathematician, who is also not something of a poet, can never be a complete mathematician."

One finds, in the West, many mathematicians who have a flair for writing poetry. Omar Khayyām, famous for Rubāis[2] and sitting under a tree with wine and women, was primarily a mathematician. In the same way, we find in Bhāskarācārya an extraordinary combination of

---

[1] Bhāskara I was a great Astronomer of the seventh century.

[2] Quatrains.

a poet and mathematician. His book amply demonstrates the qualities of head and heart.

Bhāskarācārya was born in 1114 A.D. He has made this clear in the fourth part of the *Siddhāntaśiromaṇi* titled 'Golādhyāya' (Astronomy). In the 58th stanza in the chapter of problems he says:

रसगुणपूर्णमहीसमशकनृपसमयेऽभवन्ममोत्पत्तिः।
रसगुणवर्षेण मया सिद्धांतशिरोमणी रचितः।।

I was born in *Śaka*[1] 1036 (A.D. 1114) and I wrote *Siddhāntaśiromaṇi* in *Śaka* 1072 (A.D. 1150) at the age of 36.

Bhāskarācārya has given some information about his family background at the end of the chapter on problems in Golādhyāya. From this, it follows that he belonged to the Śāṇḍilya lineage and that he lived in Vijjalavida. In the second stanza of the following verses, he talks about his art of writing with great confidence:

आसीत्सह्यकुलाचलाश्रित-पुरे त्रैविद्यविद्वज्जने।
नानासज्जनधाम्नि विज्जलविडे शाण्डिल्यगोत्रो द्विजः।।
श्रौतस्मार्तविचारसारचतुरो निःशेषविद्यानिधिः।
साधूनामवधिर्महेश्वरकृती दैवज्ञचूडामणिः।।
तज्जस्तच्चरणारविंदयुगलप्राप्तप्रसादः सुधीः।
मुग्धोद्बोधकरं विदग्धगणकप्रीतिप्रदं प्रस्फुटं।।
एतद् व्यक्तसदुक्तियुक्तिबहुलं हेलावगम्यं विदाम्।
सिद्धांतग्रथनं कुबुद्धिमथनं चक्रे कविर्भास्करः।।

Bhāskarācārya studied all sciences under the guidance of his father Maheśvara. Maheśvara was a great astrologer. According to late Mr. S.B. Dixit, Maheśvara was born in A.D. 1078. He wrote two books *Karaṇa-Grantha* and *Jātaka-Ṭīkā-Grantha.* As Bhāskarācārya studied under the guidance of such a competent teacher, he became an expert in many branches of learning.

Location of Vijjalavida

In the first stanza quoted above, Bhāskarācārya says that he hailed from Vijjalavida. But no definite information is available regarding its

---

[1] Hindu calendar.

location. The last part of the name ('Vijjalavida') is corrupted into Bida and some think that Bhāskarācārya hailed from Bida. But Bida is 64 km far from Ahmednagar in Marathawada and is not anywhere near the Sahya mountain. Besides, there no descendant of Bhāskarācārya is traceable in Bida. Some opine that Vijjalavida is the town Bedara near Hyderabad. The reason for this goes back to the great poet Faizi of Emperor Akbar's court. At the suggestion of Akbar, Faizi translated *Līlāvatī* into Persian. Therein he says that Bhāskarācārya hailed from Bedara. But Bedara, which is 80 km far from Sholapur, does not have a single small hill even nearby. Hence, Vijjalavida is not Bedara. In Bhāskarācārya's times, a town Kalyana was ruled over by Cālukya dynasty. There is neither any reference to this dynasty in Bhāskarācārya's works nor a reference to him is made in the history of Cālukya dynasty. Is Vijjalavida the town of Vijapura? Some experts consider Bhāskarācārya to be a Vaiṣṇavite Brahmin from Karnataka. Pandit Sudhakar Dwivedi also takes him to be a Vaiṣṇavite on the basis of his frequent quotes from the *Viṣṇu Purāṇa* in the chapter 'Bhuvanakośa' of *Golādhyāya*. But then Vijapura is at least 160 km east of Sahya Mountain. Also, "Vijapura" seems to be derived from "Vidyapura" or "Vijayapura" rather than "Vijjalavida". So Vijjalavida cannot be identified with Vijapura. Further, there is an evidence that Bhāskarācārya was not from Karnataka. Nṛsiṃha in his commentary *"Vāsanāvārttika" on Līlāvatī* writes

'विज्जलविडनिवासी पवित्रितदंडकारण्यः महाराष्ट्रानामाश्रयो महेश्वरनंदनः श्रीभास्कराचार्यः।'

i.e. Bhāskarācārya was from Maharashtra. In another commentrary *Marīcikā*, Munīśvara describes Vijjalavida thus:

"सह्यकुलपर्वतान्तर्गत भूप्रदेशे महाराष्ट्रदेशान्तर्गतविदर्भापरपर्यायविराटदेशादपि निकटे गोदावर्यां नातिदूरे पंच क्रोशांतरे विज्जलविडम्।"

It seems from the above quotations that Vijjalavida is situated in Khandesha or Nasika district which is north of the river Godavari. There is another Bida near Kolhapura which is close to the Sahya Mountain but not to the north of Godavari. Bida (of Marathawada), Bedara, Vijapura are no way near the ranges of Sahya Mountain.

Munīśvara wrote his commentary in 1608 which is approximately 500 years after the composition of the *Śiromaṇi.* We consider that he must have had a reliable source of information regarding the Bhāskarācārya's abode. According to him Vijjalavida was situated in between Vidarbha and Sahya Mountain. It was in between Malegaon range of Sahya Mountain and Pitala Valley in the vicinity of the Godavari, as it was about 32 to 48 km far from this river. Even if these distances are not accurate, Vijjalavida must be near Malegaon, Calisagaon, Jalagaon. This is confirmed by the research of Dr. Bhau Daji Telanga. It appears that the present Patana which is 16 km from Calisagaon must be the former Vijjalavida. Dr. Telanga discovered the inscription on land-grant to the temple of Sarajadevi of Patana. He has shown ample proof for this in the publications of the Royal Asiatic Society.[1] There he has made references to the grants given to the Pandits for the construction of Mathas, schools etc. by the rulers of Suryavanshi Dynasty, viz. Kṛṣṇarāja, Indrarāja, Govana and Sūryadeva. Further, there is a reference to a grant given to Cāṅgadeva, a grandson of Bhāskarācārya, to run a school. Nonetheless, no commentator has ever identified Vijjalavida with Patana. Ordinarily, names of towns don't change. Besides, the evidence presented by Dr. Bhau Daji shows that Cāṅgadeva alone belonged to Patana. It is possible that Cāṅgadeva and Anantadeva might have found it more conducive to run the school at Patana rather than at Vijjalavida and so they might have left their original village for good and established themselves at Patana. They might have shifted as per the ruler's orders.

This author is of the firm opinion that further research is necessary to establish the location of Vijjalavida. Until then we may accept Dr. Bhau Daji's conclusion.

No information is, however, available with us regarding the location of Bhāskarācārya's school; his patrons in his capacity as an astrologer; or his donors. So it may be concluded that Bhāskarācārya was not attached to any ruler or patron. Names of his descendants for nine generations are known from Copper Plate Inscriptions:

---

[1] Bhau Daji: *Journal of the Royal Asiatic Society* (1865), pp. 392-418 and *Journal of the Bombay Branch of the Royal Asiatic Society,* 8 (1868), p. 231.

Trivikrama, Bhāskarabhaṭṭa, Govinda, Prabhākara, Manoratha, Maheśvara, Bhāskarā, Lakṣmīdhara, Cāṅgadeva.

Bhāskarācārya's Education

It is needless to mention that Bhāskarācārya must have been an exceptionally intelligent student. He gives his genealogy in his Algebra and Question 61 of 'Golādhyāya'. In both the places he records that he received his education from his father. In Algebra he says:

> आसीत् महेश्वर इति प्रथितः पृथिव्याम् ।
> आचार्यवर्यपदवीं विदुषां प्रपन्नः ॥
> लब्धावबोधकलिकां तत एव चक्रे ।
> तज्जेन बीजगणितं लघु भास्करेण ॥

From the above it is clear that Bhāskarācāya's father was his teacher. Scholars had bestowed on him the title of "Ācārya" (Preceptor). He says that he learnt Algebra also from his father and mastered it sufficiently to write a book on it. He was well versed in Poetry, Grammar, Mathematics, Astronomy etc. and kept himself up to date. In stanza 279 of the *Līlāvatī* we learn about the subjects one had to master to earn the title 'Ācārya'. Gaṇeśa Daivajña calls Bhāskarācārya "Gaṇakacakracūḍāmaṇi" (A Jewel among mathematicians). Bhāskarācārya excelled in teaching Mathematics and Astronomy. It is clear from the Copper Plate Inscriptions of the time that the scholars hesitated to debate with Bhāskarācārya's students.

Works

Bhāskarācārya wrote his *Siddhāntaśiromaṇi* when he was 36. It is in four parts: (1) *Līlāvatī,* (2) *Algebra,* (3) *Planetary motions,* and (4) *Astronomy.* The *Līlāvatī* mainly deals with Arithmetic but also contains Geometry, Trigonometry and Algebra. Those days (as it is true even today) both educated and uneducated were interested in future events, and naturally every Astronomer had to study Astrology. For this purpose it was necessary to master Planetary Motions and Astronomy and the *Līlāvatī* provided prerequisites for this study. Bhāskarācārya wrote his *Līlāvatī* by selecting good parts from Śrīdharācārya's *Triśatikā* and Mahāvīrācārya's *Gaṇitasārasaṁgraha* and adding material of his own. All the prerequisites for the study of Astronomy are in the *Līlāvatī*. In order to solve problems concerning periods in Planetary Motions, Bhāskarācārya has added solutions of what are known as Diophantine Equations.

Bhāskarācārya's Algebra too is an excellent book on which several commentaries have been written in several languages. A well-known commentary on Algebra the *Navāṅkura* was written in 1612 by Kṛṣṇadaivajña who was an Astrologer in the court of the Emperor Jehangir. In 1634 a Persian translation of Algebra was made by Ataulla Rasidi, an Astrologer in the court of Shahjahan. Strachi translated it into English in 1813 and Khanapur Shastri rendered it into Marāṭhī in 1897. Pandit Sudhakar Dvivedi wrote a Sanskrit commentary on Algebra wherein Diophantine Equations, Squares, Pellian equations are included.

Planetary Motions and Astronomy being more involved compared to the first two parts, neither the Eastern nor Western scholars have paid any attention. However, they are worth reading by teachers and students of Mathematical Astronomy. In his Astronomical book Bhāskarācārya maintains that the earth is stationary.[1] In his work on Planetary Motions, he considers Lunar motions, Revolutions, Eclipses and such topics which are profound and involved.

Besides these, Bhāskarācārya wrote several works: *Karaṇa-kutūhala, Sarvatobhadrayantra, Vasiṣṭhatulya, Vivāhapaṭala.* It is our misfortune that the original manuscripts are not available. Maharashtrians do not generally have an awareness of the necessity of preserving historical artifacts. However, the handwritten copies of *Līlāvatī* with commentaries which are as old as two hundred years are still available.

Valiant Bhāskarācārya

Readers will notice that *Līlāvatī* is a monumental work. According to the prevailing customs, Bhāskarācārya wrote it in verse form. To explain a scientific topic like Mathematics in a verse form is indeed a difficult task. But a still more difficult job is to infuse poetic qualities.

---

[1] It appears that Bhāskara and the ancient Indian astronomers considered earth stationary just for observations and subsequent calculations only.

But Bhāskarācārya overcame all these hurdles and wrote the famous work *Līlāvatī.* If the art of printing existed then, several editions of the *Līlāvatī* might have been published. However, despite the non-availability of the printing machines, the handwritten-copies of *Līlavatī* were used from one end of India to the other. The *Līlāvatī* replaced most other texts that were used earlier. The (whole) *Līlāvatī* reached other countries a little later, in the sixteenth century. All those who read this book were overwhelmed by Bhāskarācārya's skill. It seems Bhāskarācārya himself loved his first creation, i.e., *Līlāvatī* more than his other works. Beautiful poetic flights of *Līlāvatī* are not found in any other mathematical or scientific compositions.

The Poet Bhāskarācārya

From his works it is clear that Bhāskarācārya had mastered Grammar, especially that of Pāṇini. Also, like the poets, Bhāskarācārya was fond of alliterations and metaphors, e.g.,

वदैवं समानीय पानीयमानं ।।[1]
चयं त्रयं वयं विद्मो वदनं वद नंदन ।।[2]
योजने यो जनेशः ।।[3]
बाले मरालकुलमूलदलानि सप्त ।।[4]
बाले बालमृणालशालिनि जले केलिक्रियालालसम् ।।[5]
कांते केतकमालतीपरिमलप्राप्तैककालप्रियाहूतः भृंगः ।।[6]
दर्भीयगर्भाग्रसुतीक्ष्णबुद्धिः ।।[7]

He was also quite witty. In one problem, a wealthy person is supposed to be giving one *kavaḍi*[8] in charity to a poor man! We give further evidence of Bhāskarācārya's poetical skills by examples from Astronomy. There are vivid descriptions of seasons. It is all beautiful

---

[1] See verse 160, *Līlāvatī.*
[2] Verse 131, *ibid.*
[3] Verse 133, *ibid.*
[4] Verse 73, *ibid.*
[5] Verse 75, *ibid.*
[6] Verse 60, *ibid.*
[7] Verse 43, *ibid.*
[8] Cowrie.

poetry. The late Bālakṛṣṇa Janārdhane Moḍaka rendered into Marathi the following:

सहस्यकाले बहुसस्यशालिनीं चितामवश्यायकमौक्तिकोत्करैः ।
प्रहृष्टपुष्टाखिलगोकुलामिलां विलोक्य हृष्यन्त्यधिकं कृषीवलाः ॥
(ऋतुवर्णन ६ गोलाध्याय)

In the Hemanta (winter) season the earth looks more enchanting because of the plenty of stored grains. The drops of water, in the form of dew, appear like pearls and add to the beauty. The well-fed contented cows and bulls move about in flocks and the farmers enjoy this beautiful sight.

Teaching Ability

In the *Līlāvatī* Bhāskarācārya has displayed his various skills. While teaching Mathematics he also wanted to convey informations on religion, the *Vedas*, *Purāṇas*, Epics etc. In the stanza 76 the informations are given regarding a chariot and the fact that Karṇa, though a brother of Arjuna, was his enemy. The entire battle-scene is graphically depicted. This stanza can be compared with Kālidāsa's *Śākuntalam, Grivābhaṅgābhirāmam*. The graphic descriptions of a drove of swans; a flock of elephants; a colony of bees; the attack of a snake by a domesticated peacock; sinking of a lotus in water owing to a strong wind and so on: were to train students in the appreciation of nature and to make mathematics interesting rather than tiresome. Bhāskarācārya has seldom given proofs or derivations following the contemparory tradition. Still, he gave a large number of examples.

He was a leading Astronomer of his time and had earned the well deserved title of "Jewel among the mathematicians". Bhāskarācārya handled determination of sines of angles much more skilfully than his predecessors. Although he took earth to be stationary and the centre of the universe, it did not adversely affect his calculations. Bhāskarācārya was familiar with many results that were later discovered by Copernicus and Tayko Brahe. He also had some idea of limits which was (500 years) later discovered by Newton.

He moulded a difficult language like Sanskrit to express a scientific subject like Mathematics. By using terms from fine literature, he coined many equivalents for numbers. Verily he opened a great storehouse of technical terms. If one wants Marathi[1] equivalents for scientific terms in English, one may easily find them in Bhāskarācārya's works. These could be useful to teach Mathematics in Indian languages. Further, his *Līlāvatī*, has been used as a textbook for the last eight hundred years in India. Even now it is used in some places in some Indian provinces.

He was quite practical too as he was of the opinion that if eclipses, conjunctions etc. do not occur as given in ancient almanacs, then the almanacs must be changed. Orthodox people used to stick to false beliefs, like demons Rāhu and Ketu swallow the Moon and the Sun during eclipses. What actually happens is that at the time of Lunar Eclipse, the Moon enters the shadow of the Earth and so it becomes invisible. Bhāskarācārya knew this fact but found it difficult to convince the public at large. He used to say:

> Well, Rāhu is not a demon. It is the shadow of the earth that makes the Moon invisible. If you still believe in Rāhu, at least say that he entered the shadow of the Earth and then swallowed the Moon.

This is how he found a way to reconcile the old and the new discoveries. Though a practical scholar, he was not to compromise with his principles. There was a great Indian Astronomer Lalla (A.D. 768) who wrote a textbook on Mathematics which was used until Bhāskarācārya's time. That text contained some incorrect propositions and formulae. In his book on Astronomy Bhāskarācārya corrected some of Lalla's formulae like the correction of the formula for the measurement of the surface area of a sphere. He has given this in *Līlāvatī* too. Though claimed himself to be a "student of Brahmagupta" he never hesitated to criticize the latter's long or wrong or tedious methods.[2] Sometimes he also criticized Āryabhaṭa.

The late Mr. Dixit conjectured that Bhāskarācārya might have visited Baghdada as some anonymous Indian astrologers had visited Turkey and the middle-east as honourable guests according to some Ara-

[1] Indeed all other languages derived from Sanskrit.

[2] See stanza 185b, *Līlāvatī*

bian works. If this view is acceptable, then there should have been some reference to his works to the Arabian mathematicians. Hence, the above view that Bhāskarācārya had visited the middle-east is not very convincing.

Social Conditions

It is rather difficult to draw a picture of the social conditions of the Bhāskarācārya's period by way of a book on Arithmetic. In Bhāskarācārya's times, scholarship was assessed by the study of ten works. All trade was in the hands of Lamānās. The wealthy wore garments made of silk imported from China. Young women were sold as slaves. Scholars were patronized by the rulers. Travel at night was risky. Niṣka, a silver coin, was the principal currency. Peacocks were common pets. There was abundance of lotus flowers, bees, elephants and birds. Śiva, Viṣṇu, and Gaṇeśa were universally worshipped. The *Rāmāyaṇa* and *Mahābhārata* were studied regularly. On the whole, there was prosperity. Gold was expensive but there were plenty of food items. Astronomers were more involved with Astrology than its mathematical applications. Although the scholars were generally patronised by rulers, Bhāskarācārya seems to be an exception. He was hale and hearty, and died in 1193 at the age of 79.

After Bhāskarācārya we don't find well-known learned (traditional) mathematicians in Maharashtra. Though, most students of Indian mathematics merely studied Bhāskarācārya's books still no research was done until seventeenth century.

Though a great mathematician, he did not know decimal fractions; which were discovered in Italy in the sixteenth century. It is surprising that Bhāskarācārya did not discover Newton's Binomial Theorem (in its general form). Had he discovered logarithms, determination of chords would have been considerably simplified. He knew that the earth attracted other bodies but he missed Force = Mass × Acceleration. For the computation of daily motion of planets he introduced the concept of instantaneous velocity by dividing the day into large number of small intervals. The formula $\delta(\sin\theta) = \cos\theta\,\delta\theta$ is implicitly available in *Śiromaṇi*. He was aware that the differential coefficient vanishes at an extreme value of the function. He appears to be the first mathematician who could perceive the ideas of differential calculus 500 years before Newton and Leibniz. He had little knowledge of ellipse, parabola or hyperbola. Had he proved theorems in Algebra and Geometry, he would have achieved a greater status as a mathematician. Nonetheless, his achievements put him on the top of the list of mathematicians.

We are fortunate to have had such a great mathematician in India. But we are so negligent that we forgot even to celebrate the 800th anniversary of the composition of his *Siddhāntaśiromaṇi*. Let us pledge to remember to celebrate Bhāskarācārya's eight hundredth death anniversary in 1993.

Līlāvatī
(Pāṭī Mathematics)

Praise of Lord Gaṇeśa

प्रीतिं भक्तजनस्य यो जनयते विघ्नं विनिघ्नन्
स्मृतस्तं वृंदारकवृंदवंदितपदं नत्वा मतंगाननम्।
पार्टी सद्गणितस्य वच्मि चतुरप्रीतिप्रदां प्रस्फुटाम्
संक्षिप्ताक्षरकोमलामलपदैर्लालित्यलीलावतीम्।। I ।।

[First] I offer my salutations to the elephant-faced Lord who creates love in His devotees, by remembering whom all obstacles are destroyed and whose feet are revered by the community of gods. Here I give methods of slate mathematics — *Līlāvatī* — which is loved by discriminating people because of its clarity, brevity as well as its literary flavour.

CHAPTER 1

Definitions and Tables

वराटकानां दशकद्वयं यत् सा काकिणी ताश्च पणश्चतस्रः।
ते षोडश द्रम्म इहावगम्यो, द्रम्मैस्तथा षोडशभिश्च निष्कः।। II ।।

Twenty *varāṭakas* make one *kākiṇī.* Four *kākiṇīs* make one *paṇa.* Sixteen *paṇas* make one *dramma.* And sixteen *drammas* make one *niṣka.*

Comment: Bhāskarācārya gives the following table concerning the coins used in the eleventh and twelfth centuries:

20 *kavaḍis* (Cowries) = 1 *kākiṇī* (*davaḍī*)
4 *kākiṇīs* = 1 *paṇa* (*paisā*)
16 *paṇas* (*paisās*) = 1 *dramma*
16 *dramma* = 1 *niṣka.*

Until the twentieth century, the above coins were more or less used in India. During the British regime coins *pai, ruka* and *dhabu paisa* were in circulation. Such coins were also in use during the rule of the Peshawas. In 1910 it was possible to get 40 Cowries for one Paisa and they could be used to procure sundries such as hot peppers and coriander. The word *dāma* for price has evolved from *'dramma'*.

Measure for gold

तुल्या यवाभ्यां कथितात्र गुञ्जा वल्लस्त्रिगुञ्जो धरणं च तेऽष्टौ।
गद्याणकस्तद्द्वयमिन्द्रतुल्यैः वल्लैस्तथैको धटकः प्रदिष्टः।। III ।।

2 *yavas* = 1 *guñjā* (*rattī*), 3 *guñjās* = 1 *valla*
8 *vallas* = 1 *dharaṇa*, 2 *dharaṇas* = 1 *gadyāṇaka*
14 *vallas* = 1 *dhaṭaka.*

Comment: This verse III gives various measures for weighing silver or gold. A barley-corn (with its outer shell) is called a *'yava'*. Nowadays gold and silver are weighed in gms and so *guñjās* are not in use. *Guñjā* is a fruit with black, red or blackish red colour. It is shaped like a green lentil but twice its size. *Guñjā*, being tough and durable, has been in use since ancient times. Village goldsmiths use *Guñjās* for weighing silver and gold even now. One *guñjā* weighs $1\frac{5}{16}$ troy grain. Presently, *gadyāṇa* or *gadyāṇaka* and *dhaṭaka* are not in use. But *valla* or *vāla* is still in vogue. One *dhaṭaka* = 42 *guñjās;* 1 *gadyāṇaka* = 48 *guñjās.*

दशार्धगुंजं प्रवदन्ति माषं माषाव्हयैः षोडशभिश्च कर्षम्।
कर्षैश्चतुर्भिश्च पलं तुलाज्ञाः कर्षं सुवर्णस्य सुवर्णसंज्ञम्।। IV ।।

5 *guñjās* = 1 *māṣā*, 16 *māṣās* = 1 *karṣa*
4 *karṣas* = 1 *pala.* Those who know measures for gold call a golden *karṣa "suvarṇa".*

Comment: In the times of Bhāskarācārya, the principal measure for weighing gold was a *karṣa.* 80 *guñjās* made one *karṣa.* The modern measure *tolā* equals 96 *guñjās.* 1 *tolā* = $11\frac{3}{5}$ gms.

Units of Length

यवोदरैरङ्गुलमष्टसंख्यैः हस्तोऽङ्गुलैः षड्गुणितैश्चतुर्भिः।
हस्तैश्चतुर्भिर्भवतीह दण्डः क्रोशः सहस्रद्वितयेन तेषाम्।। V ।।

8 *yavas* = 1 *aṅgula*, 24 *aṅgulas* = 1 *hasta* (forearm).

4 *hasta* = 1 *daṇḍa*, 2000 *daṇḍas* = 1 *krośa* or *kosa*.

Comment: If eight grains of barley-corn are kept bellywise close to each other, the length is one *aṅgula*, i.e., the phalanx of the fingers. There were no standard measures in India. There were many kingdoms and principalities, each having its own weights and measures. So the unit of length was not the same all over India. However, assuming that the height of a man is 3½ *hastas* (forearms), the above units could be uniformly used throughout India. Ordinarily 1 *hasta* = 20 inches, 1 inch = 2.5400 cms. So the height of a man equals 5 ft. 10 inches. A standard unit of length came into use all over India only during the British Rule. Old measures remained in the literature as well as in unimportant transactions. Now miles have given way to kilometers. One *kosa* is two miles but now both *kosa* and mile will go.

स्याद्योजनं क्रोशचतुष्टयेन तथा कराणां दशकेन वंशः।
निवर्तनं विंशतिवंशसंख्यैः क्षेत्रं चतुर्भिःश्च भुजैर्निबद्धम्।। VI ।।

1 *yojana* = 4 *kosas*, 10 *hastas* = 1 *bamboo.*

1 *nivartana* = Area of a square with sides 20 *bamboos.*

Comment: A bamboo or a pole is easily available all over for measuring lengths. The English used 1 Pole (*bamboo*) = 5½ yards. In the fifth verse, *aṅgulī* is taken as a unit of length which we can compare with foot in the English system. One *nivartana* is approximately two acres (1 acre = 0.4089 hectare).

Measures for Grains in Volume

हस्तोन्मितैः विस्तृतिदैर्ध्यपिण्डैः यद् द्वादशास्रं घनहस्तसंज्ञम्।
धान्यादिके यत् घनहस्तमानं शास्त्रोदिता मागधखारिका सा।। VII ।।

Ghanahasta (Unit Cube) is a solid which has twelve edges each of length one *hasta;* its volume is the unit. In Magadha state, quantity of grain of unit volume is called *khārī.*

Comment: *'Dvādaśāsra'* means a solid with twelve edges and not (twelve) corners as the late Khanapurkara Shastri interpreted. A cube has twelve edges but only eight corners (vertices). The measure *khārī* from Magadha must have been the standard measure for grains.

Varanasi, Allahabad, Gaya, Patna which are in southern Uttar Pradesh or Bihar, formed an influential part of ancient India. Hence measures from this part of the country were taken as standard. One *khārī* equals eight *pāyalīs* or 32 kg. *'Śāstroditā'* must be taken to mean Government order.

द्रोणस्तु खार्याः खलु षोडशांशः स्यादाढको द्रोणचतुर्थभागः।
प्रस्थश्चतुर्थांश इहाढकस्य प्रस्थांघ्रिराद्यैः कुडवः प्रदिष्टः।। VIII ।।

$\frac{1}{16}$ *khārī* = 1 *droṇa*, $\frac{1}{4}$ *droṇa* = 1 *āḍhaka*,

$\frac{1}{4}$ *āḍhaka* = 1 *prastha*, $\frac{1}{4}$ *prastha* = 1 *kuḍava*.

This is what the ancients said.

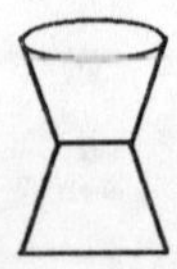

Comment: No reference is made in the verse to the shape of the measures. It appears that they were not cylindrical as they are today. They must be two frustumta of right circular cones joined together, as in the adjoining illustration. At the beginning of this century, *adholī,* Seer, Quarter Seer were used to measure grain. The words *kuḍava, āṭhave, niṭhave* are still in use in Konkana. The current Government measures are cylindrical and grain is now measured by weight.

पादोनगद्याणकतुल्यटंकैर्द्विसप्ततुल्यैः कथितोऽत्र सेरः।
मणाभिधानं खयुगैश्च सेरैर्धान्यादिमानेषु तुरुष्कसंज्ञा।। IX ।।

(The order of the digits is from right to left. Thus) *dvisapta* = 72, *kha-yuga* = 40 where *kha* denotes 0 and *yuga* denotes 4. ¾ *gadyāṇakas* = 36 *guñjās* = 1 *ṭanka,* 1 Seer = 72 *ṭankas,* 40 Seers = 1 Maund. These are Turkish measures.

Comment: Since these measures are called Turkish, this verse must have been inserted in the *Līlāvatī* later on. In Bhāskarācārya's times, there was no Muslim influence either in the north or south India. Only the north was attacked by Mahamud of Gazni. Of course, these commercial terms must have been known to businessmen but it is unlikely that they were in common use.

शेषा कालादिपरिभाषा लोकप्रसिद्धा ज्ञेया।। X ।।

Remaining measures of time etc. are well-known. E.g.: *nimiṣa,* the unit of time, measures the time for a wink (of an eye) of a wise man.
tatparā = 1/30 *nimiṣa* and *truṭi* = 1/100 *tatparā.*
18 *nimiṣas* = 1 *kāṣṭhā,* 30 *kāṣṭhās* = 1 *kalā,*
30 *kalās* = 1 *nakṣatra ghaṭikā,*
2 *ghaṭikās* = 1 *kṣaṇa,* 60 *ghaṭikās* = 1 Day.
Alternate measures of time:
1 *asu* (one complete breath) = Time taken to recite 20 long vowels.
6 *asus* = 1 *pala,* 60 *palas* = 1 *ghaṭikā.*
60 *ghaṭikās* = 1 Day, 30 Days = 1 Month, 12 Months = 1 Year.

This is the end of the chapter on terms.

CHAPTER 2

Place Values of Digits

लीलागललुलल्लोलकालव्यालविलासिने।
गणेशाय नमो नीलकमलामलकान्तये।। XI ।।

(In this verse Lord Gaṇpati is invoked again.)

I offer homage to that Lord Gaṇeśa whose neck is adorned by amusing black and most poisonous snakes and who has unctuous lustre like a blue lotus.

एकदशशतसहस्रायुतलक्षप्रयुतकोटयः क्रमशः।
अर्बुदमब्जं खर्वनिखर्वमहापद्मशंकवस्तस्मात्।।
जलधिश्चान्त्यं मध्यं परार्धमिति दशगुणोत्तरं संज्ञाः
संख्यायाः स्थानानां व्यवहारार्थं कृताः पूर्वैः।। XII ।।

Postitions of the digits from right to left are unit, ten, hundred, thousand, ten thousand, hundred thousand (*lakh*), million, ten million (*crore*), hundred million, billion (*abja*), *kharva, nikharva, mahāpadma, śanku, jaladhi, antya, madhya, parārdha.* The value of each digit on the left is ten times that on the right. This decimal system of place values was conceived by ancient scholars to make practical calculations simpler.

Comment: Indian mathematicians discovered the decimal system in which place values are assigned to digits wherein values increase in powers of ten. The Greeks and Romans used letters to represent numbers with the result that the progress of Arithmetic was very slow in Eastern Europe. In India the use of the decimal system and the use of ten symbols (for 0, 1, ..., 9) to represent any given number, made mathematical operations (addition, subtraction, etc.) easy. What Europeans call "Arabic numerals" were discovered in India and only recently some authors have started calling them "Hindu-Arabic numerals". These numerals were invented some time before 200 B.C. The current *Devanāgarī* numerals have been in use in various parts of India since A.D. 400 and the English numerals are their modified forms. Although this verse goes up to *parārdha* (10^{17}), there are terms for numbers up to 10^{140} in Sanskrit.[1]

[1] See Datta and Singh, *History of Hindu Mathematics;* and O.P. Jaggi's *Indian Astronomy and Mathematics.*

CHAPTER 3

Addition and Subtraction

अथ संकलित-व्यवकलितयोः करणसूत्रं वृत्तार्धम्।

I shall explain the method of addition and subtraction in half a stanza.

(सूत्रम्) कार्यः क्रमादुत्क्रमतोऽथवाऽङ्कयोगो यथा स्थानकमंतरं वा।। XIII ।।

Addition or subtraction should be done place-wise from right to left or *vice-versa.*

Comment: First write down the given numbers one below the other taking care to see that the digits match the places, that is, for example, digits in ten's places should be in the same vertical line. Then beginning with the unit's place add (or subtract) the digits, then move to ten's, ... In this procedure two things have been taken for granted, *viz.* (i) If the sum of the digits in the same place is greater than ten, then one has to carry forward the digit in the ten's place of the sum to the left; (ii) The number to be subtracted must be smaller than the other number. These things are not explicitly stated. Although addition and subtraction can be done from left to right also, this mehtod is not in vogue. The method is given in brief because it is clearly expected of the teacher to fill up the details. If there are more than two numbers to be subtracted from a given number, then the procedure is to subtract the numbers one by one.

अये बाले लीलावति मतिमति ब्रूहि सहितान्
द्विपञ्चद्वात्रिंशत् त्रिनवतिशताष्टादश दश।
शतोपेतानेतानयुत-वियुतांश्चापि वद मे
यदि व्यक्ते युक्तिव्यवकलनमार्गेऽसि कुशला।। XIV ।।

O! you smart girl 'Līlāvatī', if you are skilful in addition and subtraction, tell me the result when the sum of 2, 5, 32, 193, 18, 10 and 100 is subtracted from 10,000.

Comment: As shown in the adjoining figure, write the numbers to be added one below the other, keeping their proper positions (places) of digits. Then draw a horizontal line below the last number, viz. 100. Add all the digits in the unit's place to get 20, write 0 in the unit's place below the line and carry 2 to the ten's place. Now add the digits in the ten's place (along with the 2 carried over) and the sum is 16. Write 6 in the ten's place below the line and carry over 1 to the hundred's place. The sum of the digits on the hundred's place, together with the 1 carried over, is 3. Write this digit below the line in hundred's place. Thus the required sum is 360.

$$\begin{array}{r} 2 \\ 5 \\ 32 \\ 193 \\ 18 \\ 10 \\ 100 \\ \hline 360 \end{array}$$

Now as in the next figure, write the sum 360 to be subtracted below 10,000 and draw a horizontal line below 360 and the sign - before 360. In the unit's place, 0 subtracted from 0 yields 0. In the ten's place 6 cannot be subtracted from 0 and so bring 1 from the hundred's to make it 10. 6 subtracted from 10 gives 4 and we carry 1 on the hundred's. Next in hundred's place, 4 subtracted from 10 yields 6 with 1 carried over. Finally, 1 subtracted from 10 in the thousand's place, gives 9. The required answer is 9640. This explanation is to be taught by a teacher to the pupils.

$$\begin{array}{r} 10000 \\ -360 \\ \hline 9640 \end{array}$$

CHAPTER 4

Methods of Multiplication

अथ गुणने करणसूत्रं सार्द्धवृत्तद्वयम्—

Now I tell the method of multiplication in two and half stanze—

गुण्यांत्यमंकं गुणकेन हन्यादुत्सारितेनैवमुपान्त्यमादीन्।
गुण्यस्त्वधोऽधोगुणखण्डतुल्यस्तैः खण्डकैः संगुणितो युतो वा।। XV।।
भक्तो गुणः शुध्यति येन तेन लब्ध्या च गुण्यो गुणितः फलं वा।
द्विधा भवेद्रूपविभाग एवं स्थानैः पृथग्वा गुणितः समेतः।। XVI ।।
इष्टोनयुक्तेन गुणेन निघ्नोऽभीष्टघ्नगुण्यान्वितवर्जितो वा।। XVII ।।

Direct Mehtod

First multiply the digit in the unit's place of the multiplicand by the multiplier, then the digit in the ten's place and so on up to the last digit on the extreme left.

Comment: This method can be used when the multiplier is a small number and the user has memorized the multiplication tables. Of course, if the product at any stage consists of two or more digits, one has to carry over the part without the digit in the unit's place. There is no reference to this carrying over in the above rhymes; it is to be understood.

Split Method

Split the multiplier into two convenient parts, multiply the multiplicand by each of the two parts and add the results.

Comment: If the multiplier is 45, split it into 40 and 5 as it is easy to multiply by 4 and 5.

Factor Method

If the multiplier is a composite number, factor it. Then multiply by one factor and then the result by the second factor and so on. (If the multiplier is 45, first multiply by 9 and then the result by 5.)

Place Method

Multiply by each digit of the multiplier separately and write the result in each case under its proper place. Then add all the results.

Comment: This product obtained by multiplying by the digit in the unit's place is written down as it is. Then the one by the digit in the ten's place is shifted by one place to the left and written below the first one. This procedure is carried on until all the digits of the multiplier are used. Finally add all these products.

Method of Adding or Subtracting

Add any convenient number to the multiplier and multiply by the result. Then multiply by the added number and subtract this product from the previous one. Instead of addition of a convenient number one can use subtraction too.

Comment: To multiply by 45, first multiply by 50 and then by 5. Subtract the second from the first.

The above methods give the fundamentals of multiplication. To elucidate them Bhāskarācārya gives the following example.

बाले बालकुरंगलोलनयने लीलावति प्रोच्यताम्
पंचत्र्येकमिता दिवाकरगुणा अंकाः कति स्युर्यदि।
रूपस्थानविभागखंडगुणने कल्पासि कल्याणिनि
छिन्नास्तेन गुणेन ते च गुणिता अंकाः कति स्युर्वद।। XVIII ।।

O! you auspicious girl with lovable eyes of a young deer, if you have well understood the above mehtods of multiplication, what is the product of 135 and 12? Also tell me what number will you obtain if the product is divided by 12.

Comment: According to the first method we get 135 × 12 = 1620. Since 5 × 12 = 60, we write 0 in the unit's place and carry over 6. 3 × 12 = 36 to which we add 6 to get 42. We write 2 in the ten's place and carry over 4. We add 4 to 1 × 12 = 12 to get 16 which is written to the left of 2. (The explanation given by the late Khanapurkar Shastri concerning this method appears to be incorrect.)

To use the second method, we split 12 = 8 + 4. Now 135 × 8 = 1080 and 135 × 4 = 540. So 135 × 12 = 1080 + 540 = 1620.

The third method utilizes the factors 4 and 3 of 12. 135 × 3 = 405 and 405 × 4 = 1620. Of course, if the multiplier has no factors, then this method cannot be used. If the multiplicand has factors we can interchange the multiplicand and the multiplier.

```
  135
×  12
─────
  270
 135
─────
 1620
```

The fourth method uses the digits 2 and 1 in unit's and ten's place respectively. We first multiply 135 by 2 and get 270. Next we multiply by 1 and get 135 which is written by shifting it one place to the left. Then we add the two numbers to get 1620.

The fifth method uses 12 = 10 + 2. 135 × 10 = 1350 and 135 × 2 = 270. Then 135 × 12 = 1350 + 270 = 1620. Alternately 12 = 20 – 8. 135 × 20 = 2700 and 135 × 8 = 1080. 135 × 12 = 2700 – 1080 = 1620.

In stanza XVIII "the number formed by 5, 3 and 1" means 135 since the digits are written from right to left.[1] The multiplier is "*Divākara*" which means the Sun representing 12. In this stanza, there are some vocatives used to address *Līlāvatī* and we conclude that Bhāskarācārya must have been teaching a female student.

---

[1] 'अंकानां वामतो गतिः' = "Digits move in reverse order".

CHAPTER 5

Division

भागहारे करणसूत्रं वृत्तम्।

Division is explained in the following stanza–

भाज्याद्धरः शुद्ध्यति यद्गुणः स्यादन्त्यात्फलं तत्खलु भागहारे।
समेन केनाप्यपवर्त्य हारभाज्यौ भवेद्वा सति संभवे तु।। XIX।।

Find the largest integer whose product with the divisor can be subtracted from the extreme left hand digit(s) of the dividend. This integer is the first digit of the quotient. If the divisor and the dividend have a common factor, then the common factor can be cancelled and the division is carried out with the remaining factors.

Comment: To explain the process of division, we consider the example at the end of stanza XVIII: i.e., divide 1620 by 12. The first method is:

```
12) 1620 (135
    12
    ——
     42
     36
     ——
      60
      60
      ——
      00
```

12 can be subtracted from 16 once only and so the first digit of the quotient is 1 and the remainder is 4. We then have to divide 42 by 12 and 12 × 3 = 36 can be subtracted from 42. So the next digit of the quotient is 3. Now the remainder is 60 which is 12 × 5. So the last digit of the quotient is 5 and the division is complete. The quotient is 135 and remainder 0.

In the second method we use the factors 3 and 4 of 12 and 1620. Dividing 1620 by 3 we get 540. And dividing 540 by 4 we get 135.

```
3) 1620 (540                4) 540 (135
   15                          4
   ——                          ——
    12                         14
    12          and            12
   ————                        ——
   000                          20
    00                          20
   ————                         ——
    00                          00
```

In the above example the remainder is 0. If the remainder is not zero and we follow the second method, we may end up with different remainders. For example:

In $\frac{427}{28}$ the quotient is 15 and the remainder 7,

and $\frac{61}{4}$ the quotient is 15 and the remainder 1.

CHAPTER 6

Methods of Finding Squares

First Method

समद्विघातः कृतिरुच्यतेऽथ स्थाप्योन्त्यवर्गो द्विगुणान्त्यनिघ्नाः।
स्वस्वोपरिष्ठाच्च तथाऽपरेऽङ्कास्त्यक्त्वान्त्यमुत्सार्य पुनश्च राशिम्।। XX।।

The product of a number with itself is called its square. To square a number use the following procedure: First write the square of the extreme left-hand digit on its top. Then multiply the next [i.e., second] digit by the double of the first digit and write the result on the top. Next, multiply the third digit by the double of the first digit and write the result on the top. In this way arrive at the unit's place. Next cross the first digit and shift the number so formed one place to the right. Then repeat the same procedure. Finally add all the products written at the top and the sum is the required square.

Second and Third Methods

खण्डद्वयस्याभिहतिर्द्विनिघ्नी तत्खण्डवर्गैक्ययुता कृतिर्वा।
इष्टोनयुग्राशिवधः कृतिः स्यादिष्टस्य वर्गेण समन्वितो वा।। XXI ।।

Second Method

Split the given number into two parts. To the sum of the squares of the two parts add twice the product of the two parts. The result is the square.

Third Method

Add and subtract a suitable number from the given number. Take the product of the two numbers thus obtained and add the square of the suitable number chosen above. The result is the required square.

We now use the above-mentioned three methods and give the algebraic theory behind them.

सखे नवानां च चतुर्दशानां ब्रूहि त्रिहीनस्य शतत्रयस्य।
पंचोत्तरस्याप्ययुतस्य वर्गं जानासि चेद्वर्गविचारमार्गम्।। XXII ।।

O! Friend, if you know the method of finding squares, find the squares of 9, 14, 297 and 10005.

	8	8	2	0	9
D				4	9
C		1	2	6	
		8	1		
B		2	8		
	3	6			
	4				
A	2	9	7		
X		~~2~~	9	7	
Y			~~2~~	~~9~~	7

Comment: We find the squares of 297 and 10005 only.

First Method: In row A write 297. We make five vertical columns to facilitate the work. Square of 2 is 4 which is written above 2. Twice 2 is 4 and its product with 9 is 36 which is written above 4, taking care that 6 is above 9. Then we multiply 7 by 4 and write 28 above 36 with 8 above 7. We then shift 297 to the right and cross out 2 as in the row X. We repeat the procedure: square of 9 is 81 written above the line B and twice 9 multiplied by 7 = 126 written above 81 with 6 directly above 7. Then we shift 297 again, cross out 2 and 9 as in row Y. Square of 7 is 49 written above line C. Add all the products above the row A and the sum is 88209.

Some writers take the sums of numbers between A and B which is 788, between B and C which is 936 and C and D which is 49 and add thus:

$$\begin{array}{r} 788 \\ 936 \\ 49 \\ \hline 88209 \end{array}$$

Here 2nd number is shifted by one place and 3rd by two places.

The late Khanapurkar Shastri and Pandit Hariprasada Bhagiratha have not explained this point properly.

This example can be solved by other methods:

Second Method: $297 = 290 + 7$, $(290)^2 = 84100$, $(7)^2 = 49$ and $2 \times 290 \times 7 = 4060$. The sum of these three: $84100 + 49 + 4060 = 88209$.

Third Method: We choose 3 since $297 + 3 = 300$. $297 - 3 = 294$. Then

$$(297)^2 = (300) \times (294) + (3)^2$$
$$= 88200 + 9 = 88209.$$

The first method uses the formula:

$(a + b + c)^2 = a^2 + 2ab + 2ac + b^2 + 2bc + c^2$.

In the above examples these are 4, 36, 28, 81, 126 and 49, (and with proper place values 40000, 36000, 2800, 8100, 1260 and 49). It is interesting to note that the square of 97 is $8100 + 1260 + 49 = 9409$. Thus we get $7^2 = 49$, $97^2 = 9409$ and $297^2 = 88209$.

The formula used in the second method is $(a+b)^2 = a^2 + 2ab + b^2$ and the third method uses $a^2 = (a + b)(a - b) + b^2$. Of course, the second and third methods are not so useful when big numbers are involved. In the first method care should be taken to write the products in proper places.

By the second method,

$$(10005)^2 = (10000 + 5)^2$$
$$= 100000000$$
$$+ \ 100000$$
$$+ \ 25$$
$$= 100100025.$$

We also show the first method which is self-explanatory.

F	1	0	0	1	0	0	0	2	5
								2	5
E							0	0	
							0		
D					0	0	0		
					0	0			
					0				
C			0	0	0	0			
			0	0	0				
			0	0					
			0						
B				1	0				
	0	0	0	0					
	0	0	0						
	0	0							
	1								
A	1	0	0	0	5				
P		~~1~~	0	0	0	5			
Q			~~1~~	~~0~~	0	0	5		
R				~~1~~	~~0~~	~~0~~	0	5	
S					~~1~~	~~0~~	~~0~~	~~0~~	5

CHAPTER 7

Square Root

त्यक्त्वान्त्याद्विषमात् कृतिं द्विगुणयेन्मूलं समे तद्धृते
त्यक्त्वा लब्धकृतिं तदाद्यविषमाल्लब्धं द्विनिघ्नं न्यसेत्।
पंक्त्या पंक्तिहृते समेऽन्त्यविषमात्त्यक्त्वाप्तवर्गं फलम्
पंक्त्यां तद्द्विगुणं न्यसेदिति मुहुः पंक्तेर्दलं स्यात्पदम्।। XXIII।।

(Starting from the unit's place, mark alternately vertical and horizontal bars above the digits so that the given number is divided into groups of two digits each with the possible exception of the extreme left group. The extreme left group will contain either one digit or two digits and will have a vertical bar on its top or on the right digit respectively.) From the group on the extreme left, deduct the highest possible square of a_1 (say). Then write $2a_1$ in the neighbouring column; this is called *paṅkti* (row). To the right of the number obtained from the above subtraction, write the digit from the next group with a horizontal bar. Now divide the number so obtained by $2a_1$; this quotient a_2 should not be more than 9. Now write $2a_2$ below $2a_1$ after shifting it one place to the right and add. The result is the second *paṅkti.* Write the next digit to the right of the remainder so obtained and from that subtract the square of the second quotient a_2. Now to the right of the remainder so obtained, write the next digit and divide this by the second *paṅkti*. This gives the third digit of the required square root. Now twice the third digit of the square root should be added on to the second *paṅkti* after shifting it by one place to the right. The result is the third *paṅkti*. Then write the next digit of the given number to the right of the remainder and subtract from it the square of the third digit of the square root. Repeat this process. The result is the required square root.

मूलं चतुर्णां च तथा नवानां पूर्वं कृतानां च सखे कृतीनाम्।
पृथक्-पृथग्वर्गपदानि विद्धि बुद्धेर्विवृद्धिर्यदि तेऽत्र जाता।। XXIV ।।

My friend! If you really have understood the method, then find the square root of 4, 9 and of the squares obtained earlier *viz.* 81, 196, 88209, 100100025.

Comment: We'll find the square root of 88209 by Bhāskarācārya's method. The first thing is to make horizontal and vertical bars $\dot{8}\bar{8}\dot{2}\bar{0}\dot{9}$. From its first group, *viz.* 8 subtract the highest possible square which is 4. We get the first remainder 4 = 8′ – 4. Now write 8 (from the given number) to the right of the remainder 4 to get 48. 2 × 2 = 4 is the first *paṅkti*.

$\dot{8}\bar{8}\dot{2}\bar{0}\dot{9}$ (2
4
4) 48 (9
36
122
81
58) 410 (7
406
049
49
00

root	*paṅkti*	
2	4	1st
9	18	
	58	2nd
7	14	
	594	3rd
	÷ 2	
297	297	

Divide 48 by 4 and see that the highest one digit quotient does not exceed 9. Here, the quotient is 9. Write this 9 below 2 in the root column. In the same horizontal line write 2 × 9 = 18 with 1 below 4. Add the two to get 58 which is the second *paṅkti*. Then subtract 36 = 9 × 4 from 48 to get 12. To its right write the next digit 2 and we get 122. From this subtract the square of 9 to get 41. To the right of 41 write 0, the next digit from the given number. Divide 410 by the second *paṅkti* *viz.* 58 and get 7 as the quotient and 4 as remainder. Next we write this number 7 in the root column and 7 × 2 = 14 to its right with 1 below 8. Add the two to get 594 which is the third *paṅkti*. Write the last digit 9 of the given number to the right of 4 to get 49. From this subtract $7^2 = 49$ to get the remainder 0. The required square root is the number obtained by writing the digits from the root column in the order in which we derived them. Thus it is 297. We can get the same number as half of the third *paṅkti*.

In the above method of Bhāskarācārya, we take one digit at a time. In the current method, we take two digits at a time.

Below we tabulate the method of finding the square root of 100100025

```
       1̇0̄0̇1̄0̇0̄0̇2̄5̇ (1
       1
2)     00          (0
       00
       00
       00
20)    001         (0
        00
         10
          0
200)    100        (0
        000
        1000
        0000
2000)   10002      (5
        10000
            25
            25
            00
```

root	paṅkti	
1	2	1st
0	00	
	20	2nd
0	000	
	200	3rd
0	000	
	2000	4th
5	10	
10005	20010	

CHAPTER 8

Method to Find the Cube

समत्रिघातश्च घनः प्रदिष्टः स्थाप्यो घनोन्त्यस्य ततोऽन्त्यवर्गः।
आदित्रिनिघ्नस्तत आदिवर्गस्त्र्यन्त्याहतोऽथादिघनश्च सर्वे ।। XXV।।

स्थानान्तरत्वेन युतो घनः स्यात्प्रकल्प्य तत्खंडयुगं ततोऽन्त्यम्।
एवं मुहुर्वर्गघनप्रसिद्धावाद्यंकतो वा विधिरेष कार्यः।। XXVI।।

First Method

Cube of a given number is its product with itself thrice over. If we want to find the cube of a number of two digits, say, $10a + b$, write a^3 first. Below this, write $3a^2b$ by shifting this result one place to the right. Below this write $3ab^2$ after shifting it one place to the right. Below this write b^3 after shifting it one place to the right. Add all the results, and the result is the cube. This procedure can be modified by starting from b but then each time the shifting should be made to the left. If there are more than two digits, then find the cube of the two digits at the extreme left and continue with the procedure given above.

खण्डाभ्यां वा हतो राशिस्त्रिघ्नः खंडघनैक्ययुक्।
वर्गमूलघनः स्वघ्नो वर्गराशेर्घनो भवेत्।। XXVII ।।

Second Method

Or split the given number into two parts. Multiply their product by 3 times their sum. Add this to the sum of the cubes of the two parts and we get the required cube.

Another Method

The cube of a given number is the square of the cube of square root of the given number.

Comment: The mehtod given in stanza XXV depends on the algebraic identity:

$$(a + b)^3 = a^3 + 3a^2b + 3ab^2 + b^3.$$

This further leads to

$$(a + b + c)^3 = (a + b)^3 + 3(a + b)^2c + 3(a + b)c^2 + c^3$$

which can be conveniently used. The methods given in stanza XXVII depend upon the identities

$$(a + b)^3 = a^3 + b^3 + 3ab\,(a + b)$$

and $$(a^2)^3 = (a^3)^2.$$

नवघनं त्रिघनस्य घनं तथा कथय पंचघनस्य घनं च मे।
घनपदं च ततोऽपि घनात्सखे यदि घनेऽस्ति घना भवतो मतिः।। XXVIII ।।

Friend, if your comprehension of cubing is deep, tell the cubes of 9, 27 and 125.

Comment: $(9)^3 = 9 \times 9 \times 9 = 81 \times 9 = 729$. We'll find $(27)^3$ by the first two methods.

First Method

$$
\begin{array}{rcrr}
(2)^3 & = & 8 & 8 \\
3(2)^2 7 & = & 84 & 84 \\
3(2)\,(7)^2 & = & 294 & 294 \\
(7)^3 & = & 343 & 343 \\
\hline
 & & & 19683
\end{array}
$$

So $(27)^3 = 19683$.

Second Method

$$
\begin{array}{lrcr}
\text{Take} & 27 & = & 20 + 7 \\
 & 20^3 & = & 8000 \\
 & 7^3 & = & 343 \\
 & 3(20)\,(7)\,(27) & = & 11340 \\
\hline
 & & & 19683
\end{array}
$$

In this method there is no shifting as in the first method.

To cube 125 we group (12) 5.

$$
\begin{array}{rcrr}
12^3 & = & 1738 & 1728 \\
3(12)^2 5 & = & 2160 & 2160 \\
3(12)\,(5)^2 & = & 900 & 900 \\
(5)^3 & = & 125 & 125 \\
\hline
 & & (125)^3 = & 1953125
\end{array}
$$

Now we apply the last method to find $(9)^3$. Evidently

$$(9)^3 = ((3)^2)^3 = ((3)^3)^2 = (27)^2 = 729.$$

CHAPTER 9

Cube Roots

आद्यं घनस्थानमथाघने द्वे पुनस्तथान्त्याद्घनतो विशोध्य।
घनं पृथक्स्थं पदमस्य कृत्या त्रिघ्न्या तदाद्यं विभजेत् फलं तु ।। XXIX।।

पंक्त्यां न्यसेत्तत्कृतिमन्त्यनिघ्नीं त्रिघ्नीं त्यजेत्तत्प्रथमात्फलस्य।
घनं तदाद्यात् घनमूलमेवं पंक्ति भवेदेवमतः पुनश्च।। XXX।।

Draw a vertical bar above the digit in the unit's place of the number whose cube root is wanted. Then put horizontal bars on the two digits to its left, vertical bar on the next and repeat until the extreme left-hand digit is reached.

From the extreme left-hand section, deduct the highest cube possible and write to the left side the number (a) whose cube was subtracted. Write to the right of the remainder, the first digit of the next section to get new sub-dividend. Now divide by $3a^2$ and write the quotient b next to a. Then write the next digit from the section to the right of the remainder obtained above. The next divisor is $3ab^2$. In the next step take b^3 as the divisor. Continue this procedure till the digits in the given number are exhausted.

Comment: Although a similar method was used in later books in Arithmetic, there are some dissimilarities. Bhāskarācārya takes one digit at a time whereas others take groups of three. Below we find the cube root of 19683.

Root	Paṅkti	$\bar{1}\dot{9}\bar{6}\bar{8}\dot{3}$
2		8
		116
	$3(2)^2(7) = 84$	84
7		328
	$3(2)(7)^2 = 294$	294
		343
	$7^3 = 343$	343
		000

First we put the bars like $\bar{1}\dot{9}\bar{6}\bar{8}\dot{3}$. First group is 19 and the other 683. Subtract $2^3 = 8$ from 19. The remainder is 11 and we write 6 next to it to get 116. The new divisor is $3(2)^2 = 12$. We can go up to 9 but further subtractions are not possible and so we take 7 as the quotient. Write this in the root column. $3(2)^2(7) = 84$ and $116 - 84 = 32$. To the right of 32 we write 8, the next digit. The new sub-divisor is $3(2)(7)^2 = 294$ which is subtracted from 328 to give 34. We write 3 to the right of 34 and $343 = 7^3$ which when subtracted from 343 gives remainder 0. The required cube root is 27.

By the same method we now find the cube root of 1953125.

The cube root is 125.

Root	Paṅkti	$\dot{1}\bar{9}\bar{5}\dot{3}\bar{1}\bar{2}\dot{5}$
1		1
	$3(1)^2 2 = 6$	09
		6
	$3(1)(2)^2 = 12$	35
		12
2	$2^3 = 8$	233
		8
	$3(12)^2(5) = 2160$	2251
		2160
	$3 \times (12)(5)^2 = 900$	00912
		900
5	$5^3 = 125$	125
		125
		000

The current textbooks on Arithmetic don't deal with square roots and cube roots. These are now found by the use of logarithms or by the tables. Such means were not available in the days of Bhāskarācārya. The above method of Bhāskarācārya is quite simple; this simplicity disappeared in later methods. This method may be extended to obtaining n^{th} root of a positive integer.

CHAPTER 10

Eight Operations on Fractions

Reduction of Fractions to Common Denominator: Simple Fraction

अन्योन्यहाराभिहतौ हरांशौ राश्योः समच्छेदविधानमेवम्।
मिथो हराभ्यामपवर्तिताभ्यां यद्वा हरांशौ सुधियात्र गुण्यौ ।। XXXI।।

A simple fraction is one whose numerator and denominator are both integers. To add two simple fractions, first make both the denominators equal. This is done by multiplying the numerator and the denominator of the first fraction by the denominator of the second and *vice-versa.* This method can be extended to more than two fractions.

Alternatively, divide the denominators of both the fractions by their common factor and then multiply the numerator and denominator of the fractions by quotients of the other. If a wise man multiplies in this way, there will be a common denominator for both.

Examples

रूपत्रयं पंचलवस्त्रिभागो योगार्थमेतान्वद तुल्यहारान्।
त्रिषष्टिभागश्च चतुर्दशांशः समच्छिदौ मित्र वियोजनार्थम्।। XXXII।।

Friend, tell me $\frac{3}{1} + \frac{1}{5} + \frac{1}{3}$ and $\frac{1}{14} - \frac{1}{63}$.

Comment: As per the first method, we multiply the numerator and the denominator of the first by $5 \times 3 = 15$, second by 3 and the third by 5.

Example 1

$$\frac{3}{1} + \frac{1}{5} + \frac{1}{3}$$
$$= \frac{45}{15} + \frac{3}{15} + \frac{5}{15} = \frac{45 + 3 + 5}{15} = \frac{53}{15}.$$

We could have added two fractions first and add the sum to the third: $\frac{3}{1} + \frac{1}{5} = \frac{15}{5} + \frac{1}{5} = \frac{16}{5}$ and

$$\frac{16}{5} + \frac{1}{3} = \frac{48}{15} + \frac{5}{15} = \frac{53}{15}.$$

Example 2

In $\frac{1}{14} - \frac{1}{63}$ we note that 14 and 63 have a common factor 7 and the quotients (after dividing by 7) are 2 and 9 respectively.

So $\frac{1}{14} - \frac{1}{63} = \frac{9}{126} - \frac{2}{126} = \frac{7}{126} = \frac{1}{18}$.

Note that L.C.M. $\{3, 5\} = 15$ and L.C.M. $\{14, 63\} = 126$.

Compound Fractions

लवा लवघ्नाश्च हरा हरघ्ना भागप्रभागेषु सवर्णनं स्यात्।। XXXIII।।

The product of several simple fractions is obtained by dividing the product of their numerators by the product of their denominators. Then common factors of the numerator and denominator are cancelled to get the fraction in its lowest terms.

Comment: This is an example of a fraction which involves 'of'. Fraction of a fraction is also a fraction.

Example

द्रम्मार्धत्रिलवद्वयस्य सुमते पादत्रयं यद्भवेत्
तत्पंचांशकषोडशांशचरणः संप्रार्थितेनार्थिने।
दत्तो येन वराटकाः कति कदर्येणार्पितास्तेन मे
ब्रूहि त्वं यदि वेत्सि वत्स गणिते जातिं प्रभागाभिधाम्।। XXXIV।।

O good-minded guy! A miser gave to a beggar $\left(\frac{1}{2}\times\frac{2}{3}\times\frac{3}{4}\times\frac{1}{5}\times\frac{1}{16}\times\frac{1}{4}\right)$th part of a *dramma.* If you know compound fractions well then, my dear child, tell me how much money the miser gave to the beggar?

Comment: The products of the numerators and the denominators are respectively 6 and 7680. So the result is

$\frac{6}{7680} = \frac{1}{1280}$ *dramma* = 1 *kavaḍī* or Cowrie.

So the miser gave the smallest coin to the beggar. This shows the sense of humour that Bhāskarācārya possessed.

Addo and Dedo Fractions

छेदघ्नरूपेषु लवा धनर्णमेकस्य भागा अधिकोनकाश्चेत्।
स्वांशाधिकोनः खलु यत्र तत्र भागानुबंधे च लवापवाहे।
तलस्थ हारेण हरं निहन्यात् स्वांशाधिकोनेन तु तेन भागान्।। XXXV।।

Method 1

An addo fraction is one which consists of an integer plus a simple fraction whereas a dedo fraction consists of an integer minus a simple fraction. To simplify such fractions, multiply the integer by the denominator of the fractional part and then add or subtract the numerator of the fraction.

Comment: $7 + \frac{2}{3}$ is an addo fraction which is nowadays written as $7\frac{2}{3}$. Similarly $7 - \frac{2}{3}$ is a dedo fraction. $7 + \frac{2}{3} = \frac{7 \times 3 + 2}{3} = \frac{22}{3}$ which is a simple fraction. In current textbooks a 'rational fraction' is defined as $\frac{a}{b}$ where a, b are integers and $b \neq 0$.

Method 2

To add a part of a given fraction to itself, multiply the denominator of the given fraction by the denominator of the 'part' and write the product so obtained in the denominator. Then add the numerator and the denominator of the 'part' and find its product with the numerator of the given fraction. This is the numerator. To subtract a part of a given fraction from itself, suitable modification is to be made.

For example, consider the problem of adding $\frac{1}{7}$ th of $\frac{1}{5}$ to $\frac{1}{5}$. The above method gives $\frac{1\,(7+1)}{35} = \frac{8}{35}$.

The current method is $\frac{1}{5} + \frac{1}{7} \cdot \frac{1}{5} = \frac{7+1}{35} = \frac{8}{35}$.

Similarly, $\frac{1}{5} - \frac{2}{7} \cdot \frac{1}{5} = \frac{1\,(7-2)}{35} = \frac{5}{35} = \frac{1}{7}$.

By way of the present method, it would be

$$\frac{1}{5} - \frac{2}{35} = \frac{7-2}{35} = \frac{5}{35} = \frac{1}{7}.$$

सांघिद्वयं त्रयं व्यंघ्रि कीदृक् ब्रूहि सवर्णितम्।
जानास्यंशानुबंधं चेत्तथा भागपवाहनम्।। XXXVI।।

अंघ्रिः स्वत्र्यंशयुक्तः स निजदलयुतः कीदृशः कीदृशौ द्वौ
त्र्यंशौ स्वाष्टांशहीनौ तदनु च रहितौ स्वत्रिभिः सप्तभागैः।।
अर्धं स्वाष्टांशहीनं नवभिरथ युतं सप्तमांशैः स्वकीयैः
कीदृक्स्यात्ब्रूहि वेत्सि त्वमिह यदि सखेंऽशानुबंधापवाहौ।। XXXVII।।

O friend! If you know how to add a part of a fraction to that fraction and how to subtract a part of a fraction from that fraction, simplify

$2+\frac{1}{4}$, $3-\frac{1}{4}$. Also simplify (i) $\frac{1}{4}+\frac{1}{3}\left(\frac{1}{4}\right)$; (ii) $\frac{1}{2}+\frac{1}{3}\left(\frac{1}{2}\right)$;

(iii) $\frac{2}{3}-\frac{1}{8}\left(\frac{2}{3}\right)$ and $\left[\frac{2}{3}-\frac{1}{8}\left(\frac{2}{3}\right)\right]-\frac{3}{7}\left[\frac{2}{3}-\frac{1}{8}\left(\frac{2}{3}\right)\right]$;

(iv) $\frac{1}{2}-\frac{1}{8}(\frac{1}{2})$ and add $\frac{9}{7}$ of this result to itself.

Comment: $2+\frac{1}{4}=\frac{9}{4}$, $3-\frac{1}{4}=\frac{12-1}{4}=\frac{11}{4}$.

(i) $\left.\begin{matrix}\frac{1}{4}\\ \frac{1}{3}\end{matrix}\right\} \therefore \frac{1\times(3+1)}{12}=\frac{4}{12}=\frac{1}{3}$.

(ii) $\left.\begin{matrix}\frac{1}{3}\\ \frac{1}{2}\end{matrix}\right\} \therefore \frac{1\times(2+1)}{6}=\frac{3}{6}=\frac{1}{2}$.

(iii) $\left.\begin{matrix}\frac{2}{3}\\ \frac{1}{8}\end{matrix}\right\} \therefore \frac{2\times(8-1)}{3\times 8}=\frac{14}{24}=\frac{7}{12}$.

$\left.\begin{matrix}\frac{7}{12}\\ \frac{3}{7}\end{matrix}\right\} \therefore \frac{7\times(7-3)}{12\times 7}=\frac{28}{84}=\frac{1}{3}$.

(iv) $\left.\begin{matrix}\frac{1}{2}\\ \frac{1}{8}\end{matrix}\right\} \therefore \frac{1\times(8-1)}{2\times 8}=\frac{7}{16}$.

$\left.\begin{matrix}\frac{7}{16}\\ \frac{9}{7}\end{matrix}\right\} \therefore \frac{7\times(7+9)}{16\times 7}=\frac{112}{112}=1$.

CHAPTER 11

Addition and Subtraction of Fractions

योगोऽन्तरं तुल्यहरांशकानां कल्प्यो हरो रूपमहारराशेः।। XXXVIII।।

First reduce them to the common denominator. If one of the terms is an integer, take 1 as its denominator and proceed. Next add (or subtract) all the numerators so formed to make up the numerator of the result. The common denominator is the denominator. Cancel common factors, if any, and the final answer is obtained.

Comment: $\frac{5}{13} + \frac{7}{3} + 4 = \frac{5}{13} + \frac{7}{3} + \frac{4}{1} = \frac{15+91+156}{39} = \frac{262}{39}$
which is also the current method.

Example

पंचांशपादत्रिलवार्धष्ठानेकीकृतान्ब्रूहि सखे ममैतान्।
एभिश्च भागैरथ वर्जितानां किं स्यात्त्रयाणां कथयाऽऽशु शेषम्।। XXXIX।।

O friend! Tell me the sum of $\frac{1}{5}, \frac{1}{4}, \frac{1}{3}, \frac{1}{2}, \frac{1}{6}$ and the result of subtracting this sum from 3.

Comment: L.C.M. of the denominators is 60.

$$\frac{1}{5}+\frac{1}{4}+\frac{1}{3}+\frac{1}{2}+\frac{1}{6}=\frac{12+15+20+30+10}{60}=\frac{87}{60}=\frac{29}{20}.$$

$$3-\frac{29}{20}=\frac{60-29}{20}=\frac{31}{20}.$$

CHAPTER 12

Multiplication of Fractions

अंशाहितश्छेदवधेन भक्ता लब्धं विभिन्ने गुणने फलं स्यात्।। XL।।

The product of a number of fractions equals the product of their numerators divided by the product of their denominators.

Comment: $\frac{1}{2} \times \frac{1}{3} \times \frac{7}{2} \times \frac{5}{6} = \frac{35}{72}$.

Example

सत्र्यंशरूपद्वितयेन निघ्नं ससप्तमांशं द्वितयं भवेत्किम्।
अर्धं त्रिभागेण हतं च विद्धि दक्षोऽसि भिन्ने गणनाविधौ चेत्।। XLI।।

O friend! If you are an expert in multiplication of fractions, tell what is the product of $2\frac{1}{3}$, and $2\frac{1}{7}$ as well as of $\frac{1}{3}$ and $\frac{1}{2}$?

Comment: $2\frac{1}{3} \times 2\frac{1}{7} = \frac{7}{3} \times \frac{15}{7} = 5$ and $\frac{1}{3} \times \frac{1}{2} = \frac{1}{6}$.

CHAPTER 13

Division of Fractions

छेदं लवं च परिवर्त्य हरस्य शेषः।
कार्योऽथ भागहरणे गुणनाविधिश्च।। XLII।।

Division of one fraction by another is equivalent to multiplication of the first by the reciprocal of the second which is obtained by interchanging the numerator and the denominator.

Example

सत्र्यंशरूपद्वितयेन पंच त्र्यंशेन षष्ठं वद मे विभज्य।
दर्भीयगर्भाग्रसुतीक्ष्णबुद्धिश्चेदस्ति ते भिन्नहृतौ समर्था।। XLIII।।

O friend, if your intellect is as sharp as the sharp point of a blade of *kuśa*[1] grass, then give me the answers to $5 \div 2\frac{1}{3}$ and $\frac{1}{6} \div \frac{1}{3}$.

---

[1] *Doa cynosuroides.*

Comment: $5 \div 2\frac{1}{3} = \frac{5}{1} \div \frac{7}{3} = \frac{5}{1} \times \frac{3}{7} = \frac{15}{7}$

and $\frac{1}{6} \div \frac{1}{3} = \frac{1}{6} \times \frac{3}{1} = \frac{1}{2}$.

CHAPTER 14

Squares, Cubes, Square Roots and Cube Roots of Fractions

वर्गे कृती घनविधौ तु घनौ विधेयौ।
हारांशयोरथ पदे च पदप्रसिद्ध्यै।। XLIV।।

To find squares, cubes, . . . of fractions, find respectively the squares, cubes, . . . of the numerators and divide them respectively by the squares, cubes, . . . of the denominators.

Comment: $\left(\frac{7}{3}\right)^2 = \frac{7 \times 7}{3 \times 3} = \frac{49}{9}$,

$$\sqrt{\frac{81}{16}} = \frac{\sqrt{81}}{\sqrt{16}} = \frac{9}{4}.$$

सार्द्धत्रयाणां कथयाशु वर्गं वर्गात्ततो वर्गपदं च मित्र।
घनं च मूलं च घनात्ततोऽपि जानासि चेत् वर्गघनौ विभिन्नौ।। XLV।।

Find out quickly, if you know how to find squares etc. of fractions, the square and the cube of $3\frac{1}{2}$ and the square root and the cube root respectively of the numbers obtained.

Comment:

$$\left(3\frac{1}{2}\right)^2 = \left(\frac{7}{2}\right)^2 = \frac{49}{4}$$

$$\sqrt{\frac{49}{4}} = \frac{\sqrt{49}}{\sqrt{4}} = \frac{7}{2}$$

$$\left(3\frac{1}{2}\right)^3 = \left(\frac{7}{2}\right)^3 = \frac{7^3}{2^3} = \frac{343}{8}$$

$$\sqrt[3]{\frac{343}{8}} = \frac{\sqrt[3]{343}}{\sqrt[3]{8}} = \frac{7}{2}.$$

Readers would have noticed that in Bhāskarācārya's times, there were no symbols for exponents nor for radicals.

CHAPTER 15

Eight Rules Concerning Zero

योगे खं क्षेपसमं वर्गादौ खं खभाजितो राशिः।
खहरः स्यात्खगुणः खं खगुणश्चिन्त्यश्च शेषविधौ।। XLVI।।

शून्ये गुणके जाते खं हारश्चेत् पुनस्तदा राशिः।
अविकृत एव ज्ञेयस्तथैव खेनोनितश्च युतः।। XLVII।।

If zero is added to a number, the result is the same number; the square etc. (i.e., square, square-root, cube, cube-root) of zero is zero; any (non-zero) number divided by zero is *khahara,* i.e., infinite; the product of a number and zero is zero.

(If in some mathematical calculations, multiplication and division by zero are likely to occur frequently then, though a number multiplied by zero is zero,) one should maintain the form of multiplicand and multiplier zero in rest of the operations (until the final operation is reached). This is because if a number is multiplied by zero and divided by zero then the result is the (former) number. (For example,

7×0 should not be written as 0 because $7 \times 0 \div 0$ is 7; and if one writes $7 \times 0 = 0$ and if this result is divided by 0 then the result (0/0) will become indeterminate, while $7 \times 0 \div 0 = 7$.) Similarly, if zero is added to or subtracted from a number then the number remains immutated.

Comment: $a + 0 = a$ for all numbers a. $0^2 = \sqrt{0} = 0^3 = \sqrt[3]{0} = 0$. For any non-zero number a, $a/0 = \infty$. $a \times 0 = 0$ for all a. For non-zero a, $\frac{a \times 0}{0} = a$. $a \pm 0 = 0$. Especially for the rule $a/0 = \infty$, Bhāskara emphasizes in his Algebra that whenever such a situation occurs, it remains immutable in form and concept both, that is, it must be written as a/0, and that any (finite) number added to it or subtracted from it will not alter its value. He points out that this type of mathematics is used in astronomical calculations. The example given by Bhāskara in *Līlāvatī* (see Stanza XLVIII) and mathematical situations occurring in his astronomy suggest that by "1/0 is infinity" he means "$1/h \to \infty$ when $h \to 0$ through positive values". However, the symbol ∞ was not in vogue in ancient Indian mathematical works. Not only this, but several astronomers and mathematicians did not agree to Bhāskara's concept of infinity. For example, Jñānarāja (*c.* A.D. 1503) says that infinity does not remain immutable when something is added to it or subtracted from it. However they used *khahara* for infinity. *Khahara* means a fraction having a non-zero number in its numerator and zero in its denominator. It seems that the purpose of this usage of this symbol was two-fold: It was a notation for infinity. The *khahara*-form (such as 5/0) was used quite safely for further calculations or reversing the process of mathematical operations as and when needed.

The following stanza from Bhāskara's Algebra describes the nature of *khahara* (infinity).

**अस्मिन्विकारः खहरे न राशावपि प्रविष्टेष्वपि निःसृतेषु।
बहुष्वपि स्याल्लयसृष्टिकालेऽनन्तेऽच्युते भूतगणेषु यद्वत्।।**

(There is no change in infinite (*khahara*) figure if some thing is added to or subtracted from the same. It is like: there is no change in infinite

Viṣṇu (Almighty) due to dissolution or creation of abounding living beings.)

Example

खं पंचयुग्भवति किं वद खस्य वर्गं मूलं घनं घनपदं खगुणाश्च पंच।
खेनोद्धृता दश च कः खगुणो निजार्धयुक्तस्त्रिभिश्च गुणितः खहृतस्त्रिषष्टिः
।। XLVIII ।।

विलोमविधिना इष्टकर्मणा वा लब्धो राशिः ।। XLVIII A ।।

Find (i) 5 + 0, (ii) 0^2, 0^3, $\sqrt{0}$, $\sqrt[3]{0}$, (iii) 0×5, (iv) $10 \div 0$. (v) A certain number is multiplied by 0 and added to half of result. If the sum so obtained is first multiplied by 3 and then divided by 0, the result is 63. Find the original number. It is obtained by 'reverse process' (cf. next chapters).

Comment: (i) 5 + 0 = 5, (ii) $0^2 = 0^3 = \sqrt{0} = \sqrt[3]{0} = 0$, (iii) $0 \times 5 = 0$, (iv) $10 \div 0 = 10/0$. (v) If the number is x,

$$\frac{(x \times 0 + \frac{1}{2} x \times 0) \times 3}{0} = 63$$

that is $$\frac{9x}{2} \times \frac{0}{0} = 63. \quad \therefore x = 14.$$

This is the answer Bhāskarācārya would expect. In modern terminology this is equivalent to $\lim\limits_{h \to 0} \frac{(x \times h + \frac{1}{2} x \times h) \times 3}{h} = 63.$

This gives x = 14.

It is unfortunate that this modern concept of limit in traditional language could not be understood properly even by traditional Indian savants. As a consequence, the theory of modern differential calculus was delayed by about 4-5 hundred years.

CHAPTER 16

Reverse Process

छेदं गुणं गुणं छेदं वर्गं मूलं पदं कृतिम्।
ऋणं स्वं स्वमृणं कुर्यात् दृश्ये राशिप्रसिद्धये।। XLIX।।
अथ स्वांशाधिकोने तु लवाढ्योनो हरो हरः
अंशस्त्वविकृतस्तत्र विलोमे शेषमुक्तवत्।। L।।

In this 'reverse process' of obtaining an unknown quantity from the known ones, the divisor should be taken as multiplier and *vice-versa,* the square as square root, addition as subtraction and *vice-versa.* Again if by adding to or subtracting from a given number its part, we should find the indicating fraction, *viz.* $\frac{N}{D+N}$ or $\frac{N}{D-N}$ as the case may be and then apply the reverse process. Of course, the given number is to be multiplied by the indicated fraction. Next, subtract or add this result to get the unknown.

Comment: This will be made clear by examples: (i) Add 10 to a certain number. If 4 is subtracted from 2/5th of the sum so obtained, the result is 12. Find the original number.

Solution: $12 + 4 = 16$. $16 \div 2/5 = 40$

$40 - 10 = 30$ Ans.

(ii) Find the number which when added to $\frac{3}{8}$th of itself yields 33.

Solution: We have to add $\frac{3}{8}$th of the number. So multiply 33 by $\frac{3}{3+8} = \frac{3}{11}$ to get $33 \times \frac{3}{11} = 9$. So the answer is $33 - 9 = 24$.

(iii) Find the number which when $\frac{3}{8}$th part of it is subtracted yields 15.

Solution: $\frac{3}{8-3} = \frac{3}{5}$.

$15 \times \frac{3}{5} = 9$.

$15 + 9 = 25$ Ans.

Example

यस्त्रिघ्नस्त्रिभिरन्वितः स्वचरणैर्भक्तस्ततः सप्तभिः
स्वत्र्यंशेन विवर्जितः स्वगुणितो हीनो द्विपंचाशता।
तन्मूलेऽष्टयुते हृते च दशभिर्जातं द्वयं ब्रूहि तं।
राशिं वेत्सि हि चंचलाक्षि विमलां बाले विलोमक्रियाम्।। LI।।

A certain number is multiplied by 3. To this product add its $\frac{3}{4}$th part. Divide the sum so obtained by 7 and then subtract $\frac{1}{3}$ of the quotient from the quotient so obtained. Subtract 52 from the square of this remainder. Add 8 to the square root of this result. Lastly, by dividing this sum by 10 we get 2. Then O you fickle-eyed girl, if you know the 'reverse process', tell me the original number.

Comment: Using the reverse process and retracing the steps we get:

$$10 \times 2 = 20$$
$$20 - 8 = 12$$
$$12^2 = 144$$
$$144 + 52 = 196$$

$$\sqrt{196} = 14$$

$$14 + 14\left(\frac{1}{3-1}\right) = 21$$

$$21 \times 7 = 147$$

$$147 - 147\left(\frac{3}{3+4}\right) = 84$$

$$84 \div 3 = 28 \text{ Ans.}$$

By Algebra, in the equation last but one above, $x - x \times \frac{3}{7} = 84$ implies $x \times \frac{(7-3)}{7} = 84$.

Hence, $x = 84 \times \frac{7}{4} = 147$.

Bhāskarācārya takes $z + z \times \frac{3}{4} = 147$. Here also $z \times \frac{7}{4} = 147$ or $z = 147 \times \frac{4}{7} = 84$.

So the required number is 84/3 = 28.

CHAPTER 17

To Find an Unknown Quantity

(Subject to certain conditions)

उद्देशकालापवदिष्टराशिः क्षुण्णो हृतोंऽशै रहितो युतो वा।
इष्टाहतं दृष्टमनेन भक्तं राशिर्भवेत्प्रोक्तमितीष्टकर्म।। LII।।

To discover the unknown number, begin with any convenient number x. Then according to the conditions given in the problem, carry on the operations such as multiplication, division etc. Next multiply x by the result given in the problem and divide this product by the number obtained above. Thus we get the unknown by the method of supposition.

Example 1

पंचघ्नः स्वत्रिभागोनो दशभक्तः समन्वितः।
राशित्र्यंशार्धपादैः स्यात्को राशिर्द्व्यूनसप्ततिः।। LIII ।।

A certain number is multiplied by 5, and $\frac{1}{3}$ of the product is subtracted from the result. This is divided by 10 and to this quotient, half, one-third and one-fourth of the original number are added. If this is 68, find the original number.

Comment: Suppose the original number is 3. Then

$3 \times 5 = 15.$

$15 - 15 \times \frac{1}{3} = 10.$

$10 \div 10 = 1.$

$$1 + 3\left(\frac{1}{2}\right) + 3\left(\frac{1}{3}\right) + 3\left(\frac{1}{4}\right) = \frac{12 + 18 + 12 + 9}{12} = \frac{51}{12} = \frac{17}{4}.$$

Since the final result is 68, the unknown number

$$= \frac{68 \times 3}{17/4} = \frac{68}{17} \times \frac{12}{1} = 48 \text{ Ans.}$$

In this method of supposition, we have to add or subtract only parts of or multiples of the original number. Otherwise the answer may not be correct as the following example shows:

When 15 is subtracted from five times a certain number and the remainder is divided by 5 we get the quotient 29. What is the number?

Suppose the original number is 10.

Then $10 \times 5 = 50$.

$50 - 15 = 35.\ \frac{35}{5} = 7.$

Since the result is 29, the unknown number $= \frac{29 \times 10}{7}$ which is not correct.

The correct answer is 32 and since 15 is not expressed as a part or multiple of 32, the method of supposition does not work. However, in this case, the correct number may be obtained by 'reverse process' (cf. the preceding chapter). In fact,

$$\frac{29 \times 5 + 15}{5} = 32.$$

Example 2

यूथार्धं सत्रिभागं वनविवरगतं कुंजराणां च दृष्टम्
षड्भागश्चैव नद्यां पिबति च सलिलं सप्तमांशेन मिश्रः।
पद्मिन्यां चाष्टमांशः स्वनवमसहितः क्रीडते सानुरागो
नागेन्द्रो हस्तिनीभिस्तिसृभिरनुगतः का भवेद्यूथसंख्या।। LIV ।।

Of a group of elephants, half and one-third of the half went into a cave. One-sixth and $\frac{1}{7}$th of one-sixth were drinking water from a river. One-eighth and $\frac{1}{9}$th of one-eighth were sporting in a pond full of lotuses. The lover king of elephants was leading three female elephants. If this was the situation, how many elephants were there in the flock?

Comment: Suppose the number is 1.
Then the last group of elephants is 4 which is $[1-\left(\frac{1}{2}+\frac{1}{2}\times\frac{1}{3}\right)-\left(\frac{1}{6}+\frac{1}{6}\times\frac{1}{7}\right)-\left(\frac{1}{8}+\frac{1}{8}\times\frac{1}{9}\right)]$th part of the group.

Now the number in the brackets is $1-\frac{2}{3}-\frac{4}{21}-\frac{5}{36}=\frac{1}{252}$.

So the number of elephants $=\frac{4\ \times\ 1}{1/252}=4\times 252=1008$ Ans.

Example 3

अमलकमलराशेस्त्र्यंशपंचांशषष्ठैः
त्रिनयनहरिसूर्या येन तुर्येण चार्या।
गुरुपदमथ षड्भिः पूजितं शेषपद्मैः।
सकलकमलसंख्यां क्षिप्रमाख्याहि तस्य।। LV ।।

From a bunch of lotuses, $\frac{1}{3}$rd are offered to Lord Śiva, $\frac{1}{5}$th to Lord Viṣṇu, $\frac{1}{6}$th to the Sun, $\frac{1}{4}$th to the goddess. The remaining 6 were offered to the *guru*. Find quickly the number of lotuses in the bunch.

Comment: Suppose the total number of lotuses is 1.
Then the number of lotuses left is

$$1-\left(\frac{1}{3}+\frac{1}{5}+\frac{1}{6}+\frac{1}{4}\right)=1-\frac{20+12+10+15}{60}=1-\frac{57}{60}=1-\frac{19}{20}=\frac{1}{20}.$$

So the total number of lotuses is $\frac{6\times 1}{1/20}=120.$

Example 4

हारस्तारस्तरुण्या निधुवनकलहे मौक्तिकानां विशीर्णो
भूमौ यातस्त्रिभागः शयनतलगतः पंचमांशोऽस्य दृष्टः।
प्राप्तः षष्ठः सुकेश्या गणक दशमकः संगृहीतः प्रियेण
दृष्टं षट्कं च सूत्रे कथय कतिपयैर्मौक्तिकैरेष हारः ।। LVI ।।

In a coital sport of a couple, the lady's pearl necklace was broken. One-third of the pearls fell on the ground, one-fifth went under the bed. The lady collected one-sixth and her lover collected one-tenth. Six pearls remained on the original thread. Find the total number of perals in the necklace.

Comment: Suppose the total number of pearls is 1.
The final number is 6.

$$\text{Remaining pearls} = 1-\left(\frac{1}{3}+\frac{1}{5}+\frac{1}{6}+\frac{1}{10}\right)$$

$$= 1-\left(\frac{20+12+10+6}{60}\right)$$

$$= 1-\frac{48}{60}=1-\frac{4}{5}=\frac{1}{5}$$

$\therefore$ the total number of pearls $=\frac{6\times 1}{1/5}=30.$

From this example it appears that amourous problems were not taboo.

Example 5

षड्भागः पाटलीषु भ्रमरनिकरतः स्वत्रिभागः कदम्बे
पादश्चूतद्रुमे च प्रदलितकुसुमे चंपके पंचमांशः।
प्रोत्फुल्लांभोजखंडे रविकरदलिते त्रिंशदंशोऽभिरम्ये
तत्रैको मत्तभृंगो भ्रमति वद सखे का भवेद् भृंगसंख्या।। LVII ।।

O friend! One-sixth of the bees in a colony entered a *pāṭali* flower, one-third went to *kadamba* tree, one-fourth flew to a mango tree and one-fifth went to a tree blooming with *campaka* flowers. One-thirtieth went to a beautiful bed of lotuses bloomed by the Sun's rays. If only one bee was roving about, how many bees were there in the colony?

Comment: Suppose the total number of bees is 1. Then the fraction which represents the remaining one bee is

$$1-\left(\frac{1}{6}+\frac{1}{3}+\frac{1}{4}+\frac{1}{5}+\frac{1}{30}\right)$$

$$= \quad 1-\frac{10+20+15+12+2}{60}$$

$$= \quad 1-\frac{59}{60}=\frac{1}{60}.$$

So the number of bees in the colony = $\frac{1}{1/60}=60.$

Example 6

स्वार्धं प्रादात्प्रयागे नवलवयुगलं योऽवशेषाच्च काश्यां
शेषांघ्रिं शुल्कहेतोः पथि दशमलवान्षट् च शेषाद्गयायाम्।
शिष्टा निष्कत्रिषष्टिर्निजगृहमनया तीर्थपान्थः प्रयातः
तस्य द्रव्यप्रमाणं वद यदि भवता शेषजातिः श्रुताऽस्ति।। LVIII ।।

A pilgrim carried a certain amount of money. He gave away half the amount (to Brahmins) at Prayaga. He spent two-ninths of the remaining amount in Kashi. One-fourth of the remainder was paid as duty. He then spent 6/10th part of the remainder in Gaya. Finally, he returned home with 63 *niṣkas*. If you know the fractional residues, find the amount he carried.

Comment: Suppose he had 1 *niṣka* to begin with. He spent $\frac{1}{2}$ in Prayaga and so $\frac{1}{2}$ is left.

In Kashi he spent $\frac{1}{2}\cdot\frac{2}{9}=\frac{1}{9}$

and he was left with $\frac{1}{2}-\frac{1}{9}=\frac{7}{18}.$

Duty paid = $\frac{7}{18} \times \frac{1}{4} = \frac{7}{72}$;

and so he has $\frac{7}{18} - \frac{7}{72} = \frac{21}{72} = \frac{7}{24}$.

Amount spent in Gaya = $\frac{7}{24} \times \frac{6}{10} = \frac{7}{40}$.

When he left Gaya he had $\frac{7}{24} - \frac{7}{40} = \frac{7}{60}$.

$\therefore$ By the method of fractional residues, he had $\frac{63}{7/60} = 9 \times 60 = 540$ *niṣkas.*

Example 6A

लोलाक्ष्या प्रियनिर्मिता वसुलवा भूषा ललाटेऽर्पिता
यच्छिष्टास्त्रिगुणाद्रिभागकलिता न्यस्ता कुचान्तः स्त्रजि।
शिष्टार्धं भुजनालयोर्मणिगणः शेषाब्धिकस्त्र्याहतः
कांच्यात्मा मणिराशिमाशु वद मे वेण्यां हि यत्षोडश।। LIX ।।

A lover gave his fiancee some jewels for making ornaments. She used one-eighth of them for an ornament for the forehead. She used 3/7th of the remaining for a necklace. Half of the remainder were used to make armlets. Three quarters of the remaining jewels, along with little tinkling bells, were used to make a belt. Finally, she put 16 jewels in her wreathed hair. Find quickly the total number of jewels.

Comment: $\frac{1}{8}$th jewels were used for an ornament for the forehead.

So $\frac{7}{8}$ remained.

Those used for the necklace $= \frac{7}{8} \times \frac{3}{7} = \frac{3}{8}$.

Balance $\frac{7}{8} - \frac{3}{8} = \frac{4}{8} = \frac{1}{2}$.

Armlets used $\frac{1}{2} \times \frac{1}{2} = \frac{1}{4}$;

balance was $\frac{1}{4}$.

Belt needed $\frac{1}{4} \times \frac{3}{4} = \frac{3}{16}$.

Balance was $\frac{1}{4} - \frac{3}{16} = \frac{1}{16}$.

$\frac{1}{16}$th of the total jewels = 16 jewels;

and so the total number of jewels were $16 \times 16 = 256$.

Example 7

पंचांशोऽलिकुलात्कदंबमगमत्त्र्यंशं शिलीन्ध्रं तयोः
विश्लेषस्त्रिगुणो मृगाक्षि कुटजं दोलायमानोऽपरः।
कान्ते केतकमालतीपरिमलप्राप्तैककालप्रिया-
दूताहूत इतस्ततो भ्रमति खे भृंगोऽलिसंख्यां वद।। LX ।।

O deer-eyed one! there was a colony of bees. One-fifth went to the *kadamba* tree, one-third to *śilīndhra* tree. $\left(\frac{1}{3} - \frac{1}{5}\right) \times 3$ were roving around *kuṭaja* tree. One bee was attracted by the fragrance of *ketakī* and *mālatī* creepers and was moving here and there. Find the total number of bees.

Comment: In the third line there is an exquisite figure of speech. It was as if the fragrances of *ketakī* and *mālatī* were indistinguishable and the bee was unable to choose. Consequently the bee went once towards *ketakī* and then towards *mālatī* and *vice-versa* like a lover being called simultaneously by two messengers of his two beloveds.

Suppose the number to be found is 1. Then

$\left[1 - \left(\frac{1}{5} + \frac{1}{3}\right) - \left(\frac{1}{3} - \frac{1}{5}\right) \times 3\right]^{\text{th}}$ part is 1, i.e. $1 - \frac{8}{15} - \frac{6}{15} = \frac{1}{15}^{\text{th}}$ part is 1. So the number of bees = 15.

CHAPTER 18

Method of Transition

योगोऽन्तरेणोनयुतोऽर्धितस्तौ राशी स्मृतौसंक्रमणाख्यमेतत्।। LXI।।

Suppose we have to find two numbers whose sum and difference are given. Then add and subtract the given numbers and divide them by 2 to get the two numbers. This is called the method of transition.

Comment: Let the two numbers be a and b. We are given $c = a + b$, and $d = a - b$.

Clearly $a = \dfrac{c + d}{2}$, $b = \dfrac{c - d}{2}$.

ययोर्योगः शतं सैकं वियोगः पंचविंशतिः।
तौ राशी वद मे वत्स वेत्सि संक्रमणं यदि।। LXII।।

O my dear child! if you know the method of transition, find two numbers whose sum is 101 and whose difference is 25.

Comment: By the above method,

$$a = \frac{101 + 25}{2} = 63; \quad b = \frac{101 - 25}{2} = 38.$$

CHAPTER 19

Square Transition

CASE - I

वर्गान्तरं राशिवियोगभक्तं योगस्ततः प्रोक्तवदेव राशी।। LXIII।।

If the difference of the squares of two numbers is divided by their difference, we get their sum. Then using the method of transition, we get the numbers.

Comment: In Algebra we have $a^2 - b^2 = (a + b)(a - b)$.

$$\therefore \frac{a^2 - b^2}{a - b} = a + b.$$

Example

राश्योर्ययोर्वियोगोऽष्टौ तत्कृत्योश्च चतुःशती।
विवरं वद तौ राशी शीघ्रं गणितकोविद।। LXIV।।

O mathematician! tell me two numbers whose difference is 8 and the difference of whose squares is 400.

Comment: $\frac{400}{8} = 50$

$a + b = 50, a - b = 8.$

$$a = \frac{50+8}{2} = 29, \quad b = \frac{50-8}{2} = 21.$$

CASE-II

इष्टकृतिरष्टगुणिता व्येका दलिता विभाजितेष्टेन।
एकः स्यादस्य कृतिर्दलिता सैका परो राशिः ।। LXV ।।

रूपं द्विगुणेष्टहृतं सेष्टं प्रथमोऽथवापरो रूपम्।
कृतियुतिवियुती व्येके वर्गौ स्यातां ययो राश्योः ।। LXVI ।।

The second line of stanza (LXVI) poses the problem: Find two numbers a, b such that $a^2 + b^2 - 1$ and $b^2 - a^2 - 1$ are both squares ($a < b$).

First Method (LXV): Take any number x (say).
Compute $a = (8x^2 - 1)/2x$.
Then compute $b = a^2/2 + 1$.

Second Method (1st line of LXVI): Take a number x .
Then the two numbers are $1/2x + x$ and 1.

Comment: *1st Method:* Of the two methods, the first one is rather tedious. Its algebric explanation is as follows:

$$a = \frac{8x^2 - 1}{2x}, \quad b = \frac{a^2}{2} + 1.$$

$$\text{Now } a^2 + b^2 - 1 = a^2 + \frac{a^4}{4} + a^2 + 1 - 1 = \frac{a^4}{4} + 2a^2$$

$$= \frac{a^2}{4}(a^2 + 8)$$

$$= \frac{(8x^2-1)^2}{16x^2}\left[\frac{64x^4 - 16x^2 + 1 + 32x^2}{4x^2}\right]$$

$$= \frac{(8x^2-1)^2(8x^2+1)^2}{64x^4} = \left[\frac{64x^2-1}{8x^2}\right]^2$$

Similarly,

$$b^2 - a^2 - 1 = \frac{a^4}{4} + a^2 + 1 - a^2 - 1 = \left(\frac{a^2}{2}\right)^2.$$

2nd Mehtod: If x is any number, then

$$a = \frac{1+2x^2}{2x}, \quad b = 1$$

$$a^2 + b^2 - 1 = \frac{(1+2x^2)^2}{4x^2} + 1 - 1 = \left(\frac{1+2x^2}{2x}\right)^2$$

$$a^2 - b^2 - 1 = \frac{1+4x^2+4x^4-8x^2}{4x^2} = \left(\frac{1-2x^2}{2x}\right)^2.$$

The second method is easier than the first.

(i) If we take x = 1, we get $a = \frac{7}{2}$, $b = \frac{57}{8}$,

$a^2 + b^2 - 1 = \left(\frac{63}{8}\right)^2$, $b^2 - a^2 - 1 = \left(\frac{49}{4}\right)^2$. Readers can check with x = 2, 3, 4.

(ii) $x = 1$, $a = \frac{3}{2}$, $b = 1$,

$$a^2 + b^2 - 1 = \frac{9}{4}, \quad a^2 - b^2 - 1 = \frac{1}{4}.$$

NOTE: We can extend the method to get $a^2 + b^2 - c^2 = d^2$ and $a^2 - b^2 - c^2 = e^2$.

Another Derivation of the First Method: Let the two numbers be b and a + 1, $(a + 1)^2 - b^2 - 1 = a^2 + 2a - b^2$ and so it is a square if we choose $2a = b^2$. Now $(a + 1)^2 + b^2 - 1 = a^2 + 2a + b^2$ must also be a square,

i.e., by choosing $a = \frac{b^2}{2}$, $\frac{b^4}{4} + 2b^2 = \frac{b^2}{4}(b^2 + 8)$ must be a square.

It is sufficient that $b^2 + 8$ is a square. So $b = 2t - \frac{1}{t}$ for some t, since

$b^2 + 8 = 4t^2 - 4 + \frac{1}{t^2} + 8 = \left(2t + \frac{1}{t}\right)^2$. Now put t = 2x.

Then $b = 4x - \frac{1}{2x} = \frac{8x^2 - 1}{2x}$ and $a + 1 = \frac{b^2}{2} + 1$ are the two numbers.

Example

राश्योर्ययोः कृतिवियोगयुती निरेके
मूलप्रदे प्रवद तौ मम मित्र यत्र।
क्लिश्यंति बीजगणिते पटवोऽपि मूढाः
षोढोक्तगूढगणितं परिभावयन्तः।। LXVII।।

Find two numbers such that whether 1 is subtracted from the sum of their squares or difference of their squares, the result is a perfect square. O friend! even intelligent mathematicians who know the six algebraic methods fumble in trying to solve this problem.

Comment: By the first method, choosing $x = \frac{1}{2}$ we have

$$a = \frac{\frac{1}{4} \times 8 - 1}{2 \times \frac{1}{2}} = 1 \text{ and } b = \frac{a^2}{2} + 1 = \frac{3}{2}.$$

By the second method, choosing x = 1,

$$a = \frac{1}{1 \times 2} + 1 = \frac{3}{2} \text{ and } b = 1.$$

Example

इष्टस्य वर्गवर्गो घनश्च तावष्टसंगुणौ प्रथमः।
सैको राशी स्यातामेवं व्यक्तेऽथवाऽव्यक्ते।। LXVIII।।

Choose any number x. Then $a = 8x^4 + 1$ and $b = 8x^3$ are the two numbers.

Comment:

$$\begin{aligned} a^2 + b^2 - 1 &= (8x^4 + 1)^2 + 64x^6 - 1 \\ &= 64x^8 + 16x^4 + 64x^6 \\ &= [4x^2 (2x^2 + 1)]^2; \\ a^2 - b^2 - 1 &= 64x^8 + 16x^4 + 1 - 64x^6 - 1 \\ &= [4x^2 (2x^2 - 1)]^2. \end{aligned}$$

A derivation of this method is given below:
Suppose the two numbers are $a + 1$ and b. Then
$(a + 1)^2 + b^2 - 1 = a^2 + 2a + b^2$ and
$(a + 1)^2 - b^2 - 1 = a^2 + 2a - b^2$ must both be perfect squares. So we put $2a = n^2$ and $b^2 = 2an$ which makes them squares. Thus $b^2 = 2an = n^3$ and $n = 4x^2$. Then $b^2 = 64x^6$, $b = 8x^3$. In the same way, $a = \frac{n^2}{2} = 8x^4$ and $\therefore\ a + 1 = 8x^4 + 1$.

पाटीसूत्रोपमं बीजं गूढमित्येव भासते।
नास्ति गूढममूढानां नैव षोढेत्यनेकधा।। LXIX।।

Some find algebraic rules as difficult as the arithmetic ones. However, many would not find any difficulty in the above six or other rules.

अस्ति त्रैराशिकं पाटी बीजं च विमला मतिः
किमज्ञातं सुबुद्धीनामतो मन्दार्थमुच्यते।। LXX।।

In arithmetic, "the rule of three" is the essence whereas in algebra, a clear intellect is necessary. Many mathematicians may grasp easily but others would require explanations.

CHAPTER 20

Quadratic Equation

गुणघ्नमूलोनयुतस्य राशेर्दृष्टस्य युक्तस्य गुणार्धकृत्या।
मूलं गुणार्धेन युतं विहीनं वर्गीकृतं प्रष्टुरभीष्टराशिः।। LXXI।।

Suppose we are given $x^2 \pm bx$ and we are required to find x^2. Add to the above $\frac{b^2}{4}$ and find the square root of the result. Next subtract or add (as the case may be) $\frac{b}{2}$ and square the result. This is x^2.

यदा लवैश्चोनयुतः स राशिरेकेन भागोनयुतेन भक्त्वा।
दृश्यं तथा मूलगुणं च ताभ्यां साध्यस्ततः प्रोक्तवदेव राशिः।। LXXII ।।

If we are given $x^2 \pm \frac{1}{n} x^2 \pm bx$, then multiply by $\frac{1}{1 \pm \frac{1}{n}}$ and proceed as before.

Comment: (LXXI): This is the standard method in quadratic equations. $x^2 \pm bx = c$ is given. Then $x^2 \pm bx + \frac{b^2}{4} = c + \frac{b^2}{4}$.

$\therefore \left((x \pm \frac{b}{2}) \right)^2 = c + \frac{b^2}{4}$. Suppose $c + \frac{b^2}{4} = p^2$.

Then $x \pm \frac{b}{2} = p$ and $x^2 = \left(p \mp \frac{b}{2} \right)^2$.

In current practice we take $ax^2 + bx + c = 0$. Bhāskarācārya takes $a = 1$ and $-c > 0$. He also takes only positive square roots, whereas we take $\sqrt{16} = \pm 4$. So in the above working $x > 0$, $p - \frac{b}{2} > 0$. Also instead of solving for x, Bhāskarācārya solves it for x^2.

(LXXII): Here the equation is $x^2 \pm \frac{1}{n} x^2 \pm bx = c$. When simplified this becomes $x^2 \pm \frac{bx}{1 \pm \frac{1}{n}} = \frac{c}{1 \pm \frac{1}{n}}$ and then (LXXI) can be used.

Example 1

बाले मरालकुलमूलदलानि सप्त
तीरे विलासभरमन्थरगान्यपश्यम्।
कुर्वच्च केलिकलहं कलहंसयुग्मम्
शेषं जले वद मरालकुलप्रमाणम्।। LXXIII ।।

There was a flock of swans on a lakeside. Seven times half the square root of the number of swans were moving about near the lake. One amourous pair of swans was playing in water. How many swans were there?

Comment: The quadratic equation is $x^2 - \frac{7}{2} x = 2$.

$$\therefore x^2 - \frac{7}{2} x + \frac{49}{16} = 2 + \frac{49}{16} = \frac{81}{16}$$

$$\therefore \left(x - \frac{7}{4} \right)^2 = \left(\frac{9}{4} \right)^2$$

$$\therefore \left(x - \frac{7}{4}\right) = \frac{9}{4}, \text{ and } x = 4, x^2 = 16.$$

Example 2

स्वपदैर्नवभिर्युक्तं स्याच्चत्वारिंशताधिकम्।
शतद्वादशकं विद्वन् कः स राशिर्निगद्यताम्।। LXXIV ।।

O learned one, find the square number x^2 such that $x^2 + 9x = 1240$.

Comment: We have to solve $x^2 + 9x = 1240$.

$$x^2 + 9x + \frac{81}{4} = 1240 + \frac{81}{4} = \frac{5041}{4}$$

$$\therefore \left(x + \frac{9}{2}\right)^2 = \left(\frac{71}{2}\right)^2$$

$$\therefore x = \frac{71}{2} - \frac{9}{2} = 31$$

$$\therefore x^2 = 961.$$

Example 3

यातं हंसकुलस्य मूलदशकं मेघागमे मानसं
प्रोड्डीय स्थलपद्मिनीवनमगादष्टांशकोऽम्भस्तटात्।
बाले बालमृणालशालिनि जले केलिक्रियालालसं
दृष्टं हंसयुगत्रयं च सकलां यूथस्य संख्यां वद।। LXXV ।।

A flock of swans contained x^2 members. As the clouds gathered, 10x went Mānasa Lake, and $\frac{1}{8}$ x^2 flew away to a garden of *Hibiscus Mutalis.* The remaining three amourous couples played about in the water. O young woman, how many swans were there in that lake full of beautiful little lotuses?

Comment: Here the equation is $x^2 - 10x - \frac{1}{8}x^2 = 6$.

$$\therefore \frac{7}{8}x^2 - 10x = 6$$

$$\therefore x^2 - \frac{80x}{7} = \frac{48}{7}$$

$$\therefore x^2 - \frac{80x}{7} + \frac{(40)^2}{49} = \frac{48}{7} + \frac{(40)^2}{49} = \frac{1936}{49}$$

$$\therefore x = \frac{44}{7} + \frac{40}{7} = 12 \text{ and } x^2 = 144 .$$

Example 4

पार्थः कर्णवधाय मार्गणगणं क्रुद्धो रणे संदधे
तस्यार्धेन निवार्य तच्छरगणं मूलैश्चतुर्भिर्हयान्।
शल्यं षड्भिरथेषुभिस्त्रिभिरपि च्छत्रं ध्वजं कार्मुकम्
चिच्छेदास्य शिरः शरेण कति ते यानर्जुनः संदधे।। LXXVI ।।

Arjuna became furious in the war and, in order to kill Karṇa, picked up some arrows (say x^2). With half of the arrows ($\frac{1}{2} x^2$), he destroyed all of Karṇa's arrows. He killed all of Karṇa's horses with four times the square root of the arrows (4x). He destroyed the spear with 6 arrows. He used one arrow each to destroy the top of the chariot, the flag, and the bow of Karṇa. Finally he cut off Karṇa's head with another arrow. How many arrows did Arjuna discharge?

Comment: Here the equation is $x^2 - \frac{x^2}{2} - 4x = 6 + 3 + 1 = 10$

$$\text{i.e., } x^2 - 8x = 20$$

$$\therefore x^2 - 8x + 16 = 20 + 16 = 36 = 6^2$$

$$\therefore x = 10, x^2 = 100.$$

Example 5

अलिकुलदलमूलं मालतीं यातमष्टौ
निखिलनवमभागाश्चालिनी भृंगमेकम्।
निशि परिमललुब्धं पद्ममध्ये निरुद्धम्
प्रतिरणति रणन्तं ब्रूहि कांतेऽलिसंख्याम्।। LXXVII ।।

From a group of black bees ($2x^2$), square root of the half (x) went to the *mālatī* tree. Again $\frac{8}{9}$th of the bees went to the *mālatī* tree. Of the remaining two, one got caught in a lotus whose fragrance captivated him; he started wailing and his beloved responded. Then, O beloved, how many bees were there?

Comment: Here the equation is $2x^2 - x - \frac{8}{9}(2x^2) = 2$, i.e.,

$$\frac{2x^2}{9} - x = 2$$

$$\therefore x^2 - \frac{9}{2}x = 9$$

$$\therefore x^2 - \frac{9}{2}x + \frac{81}{16} = 9 + \frac{81}{16} = \frac{225}{16}$$

$$\therefore \left(x - \frac{9}{4}\right)^2 = \left(\frac{15}{4}\right)^2$$

$$\therefore x = \frac{15}{4} + \frac{9}{4} = 6 \quad \therefore 2x^2 = 72.$$

Example 6

यो राशिरष्टादशभिः स्वमूलैः
राशित्रिभागेन समन्वितश्च।
जातं शतद्वादशकं तमाशु
जानीहि पाट्यां पटुताऽस्ति ते चेत्।। LXXVIII ।।

To a certain number (x^2), the 18 times of the square root of the number (*viz.,* 18x) is added. When one-third of the number $\left(viz.\ \frac{1}{3}x^2\right)$ is added to this sum, the result is 1200. If you know arithmetic well, tell the number.

Comment: Here the equation is $x^2 + 18x + \frac{1}{3}x^2 = 1200$

$$\therefore 4x^2 + 54x = 3600$$

$$\therefore x^2 + \frac{27}{2}x = 900$$

$$\therefore x^2 + \frac{27}{2}x + \frac{729}{16} = 900 + \frac{729}{16}$$

$$\therefore \left(x + \frac{27}{4}\right)^2 = \frac{15129}{16} = \left(\frac{123}{4}\right)^2$$

$$\therefore x = \frac{123}{4} - \frac{27}{4} = 24$$

$$\therefore x^2 = 576.$$

CHAPTER 21

The Rule of Three

प्रमाणमिच्छा च समानजातिः
आद्यन्तयोस्तत्फलमन्यजातिः ।
मध्ये तदिच्छाहतमाद्यहृत्स्यात्
इच्छाफलं व्यस्तविधिर्विलोमे ।। LXXIX ।।

There are three quantities involved herein. The first one on the left (a) is called *pramāṇa* (scale), the second (b) *phala* (result or fruit) and the third (c) *icchā* (desire or requisition). The answer to be found (d) is called *icchā-phala* (desired result). Here a and c must be of the same kind and b should be different from a and c. The formula is $d = \frac{b \times c}{a}$ which is of the same kind as b. In a reverse case, the inverted rule of three is applied.

Comment: (i) $\frac{a}{b} = \frac{c}{d}$ in case of direct variation. and (ii) ab = cd in case of inverse variation. So in the first case $d = \frac{b \times c}{a}$ and in the second case, $d = \frac{a \times b}{c}$.

इच्छावृद्धौ फले वृद्धिर्ह्रासे ह्रासश्च जायते।
समत्रैराशिकं तत्र ज्ञेयं गणितकोविदैः।। LXXX ।।

If, in a given situation, the desired result (d) increases (respectively decreases) as the requisition (c) increases (resp. decreases) then the simple rule of three is applied by an adept mathematician.

Example 1

कुंकुमस्य सदलं पलद्वयं निष्कसप्तमलवैस्त्रिभिर्यदि।
प्राप्यते सपदि मे वणिग्वर ब्रूहि निष्कनवकेन तत्कियत्।। LXXXI ।।

If $2\frac{1}{2}$ *(Pala)* of saffron costs $\frac{3}{7}$ N (*niṣkas*) O you expert business-man, tell me quickly what quantity of saffron can be bought for 9 *niṣkas*.

Comment: 5 *guñjās* = 1 *māṣā* and 64 *māṣās* = 1 *pala*. So 1 *pala* = 320 *guñjās*. $2\frac{1}{2}$ *pala* = 800 *guñjās*. The rule of three gives

$\frac{3}{7} : \frac{5}{2} :: 9 : d$ and thus $d = \frac{9 \times \frac{5}{2}}{3/7} = \frac{105}{2}$ *palas*. This is direct proportion, since more money buys more saffron.

NOTE: Here *kumkuma* means saffron and not the red powder that is applied to a forehead, as the late Khanapurkar Shastri takes it to mean. In 1937, two rupees bought $8\frac{1}{2}$ *tolās* of saffron. The red powder is cheap even now.

Example 2

प्रकृष्टकर्पूरपलत्रिषष्ट्या चेल्लभ्यते निष्कचतुष्कयुक्तम्।
शतं तदा द्वादशभिः सपादैः पलैः किमाचक्ष्व सखे विचिंत्य।। LXXXII ।।

63 *palas* of purified camphor is sold in the market for 104 *niṣkas.* O friend! tell me, after giving a thought, what will $12\frac{1}{4}$ *palas* fetch.

Comment: This is an example of direct proportion.

$$63 \text{ P} : 104 \text{ N} :: 12\frac{1}{4} \text{ P} : \text{d}.$$

$$\text{So d} = \frac{49}{4} \times \frac{104}{63} = \frac{182}{9} = 29\frac{2}{9} \text{ N}.$$

So the answer is 20 *niṣkas,* 3 *drammas,* 8 *paṇas,* 3 *kākiṇīs* and $11\frac{1}{9}$ *varāṭakas.*

Example 3

द्रम्मद्वयेन साष्टांशा शालितण्डुलखारिका।
लभ्या चेत् पणसप्तत्या तत्किं सपदि कथ्यताम्।। LXXXIII ।।

If $1\frac{1}{8}$ *khārikās* of rice can be bought for 2 *drammas,* then how much rice can be bought for 70 *paṇas*?

Comment: This is also an example of direct proportion.

$$32 \text{ P} : \frac{9}{8} \text{ K} :: 70 \text{ P} : \text{d}$$

$$\therefore \text{d} = \frac{70}{1} \times \frac{9}{8} \times \frac{1}{32} = \frac{315}{128} = 2\frac{59}{128} \text{ K}.$$

(Here P is for *paṇa* and K for *khārikā.*)

The rule of three perfected in India, and radiated in the eighth century. In Europe, it was held in very high esteem and called the Golden Rule.

CHAPTER 22

Inverse Proportion

इच्छावृद्धौ फले ह्रासो ह्रासे वृद्धिश्च जायते।
व्यस्तं त्रैराशिकं तत्र ज्ञेयं गणितकोविदैः।। LXXXIV।।

If, in a given situation, the desired result (d) decreases (resp. increases) as the requisition (c) increases (resp. decreases) then the inverted rule of three is applied by an adept mathematician.

Comment: Common sense (which is uncommon!) should be used to determine whether the fourth term is going to be smaller or larger than the second term. Then use the rule of three. At present unitary method has replaced the rule of three and in the West, the rule of three has become obsolete. Naturally rules of five and higher numbers have also disappeared. The authors are of the firm opinion that the rule of three is simpler than the unitary method.

जीवानां वयसो मौल्ये तौल्ये वर्णस्य हेमनि।
भागहारेच राशीनां व्यस्तं त्रैराशिकं विदुः।। LXXXV ।।

Vigorous age of a creature is inversely proportional to its price. Same is true of the relations between the desired fineness of gold and its weight, and given unit of measure and the quantity of grain.

Example 1

प्राप्नोति चेत् षोडशवत्सरा स्त्री द्वात्रिंशकं विंशतिवत्सरा किम्।
द्विधूर्वहो निष्कचतुष्कमुक्षा प्राप्नोति धूःषट्कवहस्तदा किम्।। LXXXVI ।।

If the price of a sixteen-year-old female slave is 32 *niṣkas,* find the price of a twenty-year-old one. An ox which has been yoked for two years costs 4 *niṣkas*: what will an ox which has been yoked for six years cost?

Comment: (i) Price decreases as the age increases.

So $d = \dfrac{32 \times 16}{20} = \dfrac{128}{5} = 25\dfrac{3}{5}$ N.

(ii) Here the price is $\dfrac{4 \times 2}{6} = \dfrac{4}{3} = 1\dfrac{1}{3}$ N.

Example 2

दशवर्णं सुवर्णं चेद् गद्याणकमवाप्यते।
निष्केण तिथिवर्णन्तु तदा वद कियन्मितम्।। LXXXVII ।।

One *'gadyāṇaka'* (= 48 *'guñjās'*) of 10 carat gold can be bought for 1 N. How much 15 carat gold can be bought for the same price?

Comment: Finer the gold, less can be bought for the same price. So quantity of 15 carat gold = $\dfrac{10}{15} \times \dfrac{48}{1} = 32$ *guñjās.*

Example 3

सप्ताढकेन मानेन राशौ सस्यस्य मापिते।
यदि मानशतं जातं तदा पंचाढकेन किम्।। LXXXVIII ।।

With a measure of 7A (*āḍhakas*), a certain quantity of grain measures 100 units. How many units will there be if the measure is 5A?

Comment: This is also an example of inverse proportion. The answer is $\dfrac{100 \times 7}{5} = 140$ units.

CHAPTER 23

The Rule of Five

पंचसप्तनवराशिकादिकेऽन्योन्यपक्षनयनं फलच्छिदाम्।
संविधाय बहुराशिजे वधे स्वल्पराशिवधभाजिते फलम्।। LXXXIX।।

In the case of examples on the rules of five, seven, nine, etc. keep the antecedents of all proportions in the numerator. All the other terms, except the desired result, should be kept in the denominator. The product of the numerators divided by the product of the denominators is the required result.

Example 1

मासे शतस्य यदि पञ्चकलान्तरं स्यात्
वर्षे गते भवति किं वद षोडशानाम्।
कालं तथा कथय मूलकलान्तराभ्याम्
मूलं धनं गणक कालफले विदित्वा।। XC ।।

There are three problems in this.

(1) If 100 *niṣkas* (= N) fetch 5 N interest per month, find the interest on 16 N for one year.

$$\text{Interest} = \frac{16 \times 12 \times 5}{100 \times 1} = \frac{48}{5} = 9\frac{3}{5}\,\text{N}.$$

In a tabular form, the rule of five is:

$$\left.\begin{array}{r} 100\text{ N Principal} : 16\text{ N Principal} \\ 1\text{ month} : 12\text{ months} \end{array}\right\} : : 5\text{ N interest} : x$$

(2) The above problem is altered: at the same rate as in (1), find the period to fetch $9\frac{3}{5}$ interest on 16 N.

Let the period be x months.

$$\left.\begin{array}{r} 100\text{ N} : 16\text{ N} \\ 5\text{ N} : \frac{48}{5}\text{ N} \end{array}\right\} : : 1\text{M} : x \quad \left.\begin{array}{l} \text{inverse} \\ \text{direct} \end{array}\right\}$$

In the above, the first proportion is inverse since if the principal is less, the period is longer to gain the same interest.

$$\text{So } x = \frac{100 \times 1}{16 \times 5} \times \frac{48}{5} = 4 \times 3 = 12 \text{ months.}$$

(3) Suppose we are given the period and interest and we have to find the principal (x). Here

$$\left.\begin{array}{r} 5\text{ N} : \frac{48}{5}\text{ N} \\ 1\text{M} : 12\text{ M} \end{array}\right\} : : 100\text{ N} : x \quad \left.\begin{array}{l} \text{direct} \\ \text{inverse} \end{array}\right\}$$

$$x = \frac{100 \times 48}{5 \times 5} \times \frac{1}{12} = 16\,\text{N}.$$

Example 2

सत्र्यंशमासेन शतस्य चेत्स्यात्कलान्तरं पञ्च सपञ्चमांशाः।
मासैस्त्रिभिः पञ्चलवाधिकेस्तत्सार्धद्विषष्टेः फलमुच्यतां किम्।। **XCI** ।।

If the interest on 100 for $\frac{4}{3}$ months is $5\frac{1}{5}$, what will be the interest on $62\frac{1}{2}$ for $3\frac{1}{5}$ months?

Comment: The rule of five is:

$$\left.\begin{array}{l} 100 : 62\frac{1}{2} \\ \frac{4}{3}\text{M} : \frac{16}{5}\text{N} \end{array}\right\} : : \frac{26}{5} : \text{x} \quad \left.\begin{array}{l}\text{direct}\\ \text{direct}\end{array}\right\}$$

$$\text{x} : \frac{26}{5} \times \frac{125}{2} \times \frac{16}{5} \times \frac{1}{100} \times \frac{3}{4} = \frac{39}{5} = 7\frac{4}{5}\ .$$

Example 3

विस्तारे त्रिकराः कराष्टकमिता दैर्घ्ये विचित्राश्च चेत्
रूपैरुत्कपट्टसूत्रपटिका अष्टौ लभन्ते शतम्।
दैर्घ्ये सार्धकरत्रयाऽपरपटी हस्तार्धविस्तारिणी
तादृक्किं लभते द्रुतं वद वणिग्वाणिज्यकं वेत्सि चेत्।। XCII ।।

For 100 N (*niṣkas*) 8 pieces of multi-coloured embroidered (clothing) material, each measuring 3 cubits × 8 cubits, are available. O businessman, if you are proficient in business, tell me quickly the price of a piece $3\frac{1}{2}$ cubits × $\frac{1}{2}$ cubit.

Comment: This is an example of the rule of seven.

$$\left.\begin{array}{l} 3\text{ cubits} : \frac{1}{2}\text{ cubit} \\ 8\text{ cubits} : \frac{7}{2}\text{ cubits} \\ 8 \qquad\quad : 1 \end{array}\right\} : : 100\text{ N} : \text{x} \quad \left.\begin{array}{l}\text{direct}\\ \text{direct}\\ \text{direct}\end{array}\right\}$$

$$\therefore\ \text{x} = \frac{100}{1} \times \frac{7}{2} \times \frac{1}{2} \times \frac{1}{3 \times 8 \times 8}$$
$$= \frac{175}{192}\text{N}\ .$$

Example 4

पिण्डे येऽर्कमिताङ्गुलाः किल चतुर्वर्गांगुला विस्तृतौ
पट्टा दीर्घतया चतुर्दशकरास्त्रिंशल्लभन्ते शतम्।
एता विस्तृतिपिण्डदैर्ध्यमितयो येषां चतुर्वर्जिताः
पट्टास्ते वद मे चतुर्दश सखे मूल्यं लभन्ते कियत्।। XCIII ।।

30 planks of wood, each measuring 14 cubits × 16 fingers × 12 fingers cost 100 N (*niṣkas*). Find O friend, the cost of 14 such planks whose dimensions are each 4 measures less than the former.

Comment: This is an example on the rule of nine:

$$\left.\begin{array}{l}\left.\begin{array}{ll}\text{14 cubits} & :\text{10 cubits}\\ \text{16 fingers} & :\text{12 fingers}\\ \text{12 fingers} & :\text{8 fingers}\\ 30 & :14\end{array}\right\} : : \text{100 N} : x\end{array}\right\}\text{All are direct proportions.}$$

$$\therefore\ x = \frac{10 \times 12 \times 8 \times 14 \times 100}{14 \times 16 \times 12 \times 30 \times 1} = \frac{50}{3} = 16\frac{2}{3}\ \text{N}.$$

Example 5

पट्टा ये प्रथमोदित्तप्रमित्तयो गव्यूतिमात्रे स्थिता:
तेषामानयनाय चेच्छकटिनां द्रम्माष्टकं भाटकम्।
अन्ये ये तदनंतरं निगदिता माने चतुर्वर्जिता:
तेषां किं भवतीति भाटकमित्तिर्गव्यूतिषट्के वद।। XCIV ।।

If 8 *drammas* are transport charges for carrying the planks, of the previous example, through a distance of 1 *kosa,* find out the charges for transporting the second set of planks (measuring 4 units less in each dimension) through a distance of 6 *kosas.*

Comment: This is to be solved by the rule of eleven.

$$\left.\begin{array}{l}\left.\begin{array}{ll}\text{14 cubits} & :\text{10 cubits}\\ \text{16 fingers} & :\text{12 fingers}\\ \text{12 fingers} & :\text{8 fingers}\\ 30 & :14\\ \text{1 K} & :\text{6 K}\end{array}\right\} : : \text{8 D} : x\end{array}\right\}\text{All are direct proportions.}$$

$$\therefore\ x = \frac{10 \times 12 \times 8 \times 14 \times 6 \times 8}{14 \times 16 \times 12 \times 30 \times 1} = 8\ \textit{drammas}.$$

CHAPTER 24

Rules for Barter

तथैव भांडप्रतिभांडकेऽपि विधिर्विपर्यस्य हरांश्च मूल्ये ।। XCV।।

In different kinds of goods are to be exchanged, use the rules of three, five etc. as explained earlier. However, the price and the quantity of goods are in inverse proportion.

Example

द्रम्मेण लभ्यत इहाऽऽम्रशतत्रयं चेत्
त्रिंशत्पणेन विपणौ वरदाडिमानि
आम्रैर्वदाऽशु दशभिः कति दाडिमानि
लभ्यानि तद्विनिमयेन भवन्ति मित्र।। XCVI ।।

In a market, 300 mangoes can be purchased for 1 *dramma* = 16 *paise*. However, 30 pomegranates of good quality are available for 1 *paisā*. Find out quickly how many pomegranates can be exchanged for 10 mangoes.

Comment: *First Method:* As per Bhāskarācārya's method.

$$\left.\begin{matrix}300 \text{ mangoes} : 10 \text{ mangoes} \\ 16 \text{ Paise} : 1 \text{ Paisa}\end{matrix}\right\} :: 30 : x \quad \left.\begin{matrix}\text{direct} \\ \text{inverse}\end{matrix}\right\}$$

$$\therefore x = \frac{10 \times 30 \times 16}{300 \times 1} = 16 \text{ Pomegranates.}$$

Second Method: First we determine the number of pomegranates equivalent to 300 mangoes = 16 × 30 = 480

$$300 \text{ mangoes} = 480 \text{ pomegranates}$$

$$\therefore \quad 10 \text{ mangoes} = \frac{480}{300} \times \frac{10}{1} = 16 \text{ pomegranates.}$$

CHAPTER 25

Simple Interest

प्रमाणकालेन हतं प्रमाणं विमिश्रकालेन हतं फलं च।
स्वयोगभक्ते च पृथक् स्थिते च मिश्राहते मूल कलान्तरे स्तः।। XCVII।।

(To compute simple interest and principal.) Multiply the standard principal (100) by the standard period (1 month or 1 year). Next multiply the given period by the given rate of interest. Keep the two products a, b separately. Multiply a by the amount and divide it by (a + b) to get the principal. Similarly, the amount multiplied by b and divided by (a + b) yields the interest.

Comment: A = amount, P = principal, I = interest,
R = rate of interest, Y = Period.
P_0 = standard principal (usually 100)
Y_0 = standard period (1 year or 1 month).

$$P = \frac{A \times P_0 \times Y_0}{P_0\, Y_0 + RY}$$

$$I = \frac{A \times R \times Y}{P_0 Y_0 + RY}.$$

According to the present method, first calculate the interest on Rupees (Rs.) 100 for the given period and add it to Rs. 100 to get the amount for Rs. 100. By the rule of three then find the principal and then interest.

Example

पंचकेन शतेनाब्दे मूलं स्वं सकलान्तरम्।
सहस्त्रं चेत्पृथक् तत्र वद मूल कलान्तरे।। XCVIII ।।

When the interest rate is 5% per month, the amount after one year is 1000 N (*niṣkas*). Find the principal and the interest.

Comment: The above stanza does not mention 'per month' but it seems that in those times the interest was calculated on monthly basis.

$$P = \frac{1000 \times 100 \times 1}{100 \times 1 + 12 \times 5} = 625 \text{ N}$$

$$I = \frac{1000 \times 12 \times 5}{100 \times 1 + 12 \times 5} = 375 \text{ N}$$

We could have got I = A – P = 1000 – 625 = 375 N. The interest 375 N on 625 N for one year appears high.

अथ प्रमाणैर्गुणिताः स्वकाला व्यतीतकालघ्नफलोद्धृतास्ते।
स्वयोगभक्ताश्च विमिश्रनिघ्नाः प्रयुक्तखंडानि पृथक् भवन्ति।। XCIX ।।

(If several parts of a certain principal bear different rates of interest for different periods and yet yield the same interest, to find these parts.) Take the product of the standard principal and the standard period, divide this product by the product of respective periods and rates of interest, and write these quotients separately. These quotients multiplied by the given principal and divided by the sum of the quotients written separately are the desired parts of the given principal.

Comment: Using the notation of the previous example with suffixes to denote the parts (we consider three parts):

$$P_1 = \frac{(P_1+P_2+P_3)\times\frac{100}{R_1\,Y_1}}{\frac{100}{R_1\,Y_1}+\frac{100}{R_2\,Y_2}+\frac{100}{R_3\,Y_3}} = \frac{(P_1+P_2+P_3)\times\frac{1}{R_1\,Y_1}}{\frac{1}{R_1\,Y_1}+\frac{1}{R_2\,Y_2}+\frac{1}{R_3\,Y_3}} \text{ etc.}$$

Interest in each case is:

$$\frac{P_1\,R_1\,Y_1}{100} = \frac{P_2\,R_2\,Y_2}{100} = \frac{P_3\,R_3\,Y_3}{100}$$

$$\therefore \frac{P_1}{\frac{100}{R_1\,Y_1}} = \frac{P_2}{\frac{100}{R_2\,Y_2}} = \frac{P_3}{\frac{100}{R_3\,Y_3}} = \frac{P_1 + P_2 + P_3}{\frac{100}{R_1\,Y_1}+\frac{100}{R_2\,Y_2}+\frac{100}{R_3\,Y_3}}$$

from which we get the formulae for P_1, P_2, P_3.

यत्पंचकत्रिकचतुष्कशतेन दत्तं
खंडैस्त्रिभिर्गणक निष्कशतं षडूनम्।
मासेषु सप्तदशपंचसु तुल्यमाप्तम्
खंडत्रयेऽपि हि फलं वद खंडसंख्याम्।। C ।।

94 N (*niṣkas*) were divided into three parts and were lent at 5 per cent (per month) for 7 months, at 3 per cent for 10 months, at 4 per cent for 5 months. If the three parts yield equal interest, find them.

Comment: Here $P_1 + P_2 + P_3 = 94$

$$\frac{1}{R_1\,Y_1} = \frac{1}{35}, \frac{1}{R_2\,Y_2} = \frac{1}{30}, \frac{1}{R_3\,Y_3} = \frac{1}{20}$$

$$\frac{1}{R_1\,Y_1}+\frac{1}{R_2\,Y_2}+\frac{1}{R_3\,Y_3} = \frac{1}{35}+\frac{1}{30}+\frac{1}{20} = \frac{47}{420}$$

$$P_1 = \frac{94}{1}\times\frac{420}{47}\times\frac{1}{35} = 24, P_2 = \frac{94}{1}\times\frac{420}{47}\times\frac{1}{30} = 28$$

$$P_3 = \frac{94}{1}\times\frac{420}{47}\times\frac{1}{20} = 42.$$

Now Bhāskarācārya gives a formula to compute the shares of profit when total profit and individual investments are given.

प्रक्षेपका मिश्रहता विभक्ता प्रक्षेपयोगेन पृथक् फलानि।। CI ।।

An individual's share (after business) is the individual's investment multiplied by the total output and divided by the total investment.

Comment: If a, b, c are investments and x is the output, then the shares are ax/(a + b + c), bx/(a + b + c), cx/(a + b + c) respectively.

पंचाशदेकसहिता गणकाष्टषष्टिः
पंचोनिता नवतिरादिधनानि येषाम्।
प्राप्ता विमिश्रितधनैस्त्रिशती त्रिभिस्तैः
वाणिज्यतो वद विभज्य धनानि तेषाम्।। CII ।।

Three grocers invested 51, 68, 85 N (*niṣkas*) respectively. Skillfully they increased their total assets to 300 N. Find the share of each.

Comment: Total investment = 51 + 68 + 85 = 204 N.
Their shares are:

$$\frac{51 \times 300}{204} = 75\text{ N}, \quad \frac{68 \times 300}{204} = 100\text{ N}, \quad \frac{85 \times 300}{204} = 125\text{ N}.$$

Profits are 24 N, 32 N, 40 N.

We have now a formula concerning filling up of reservoirs (pools, lakes, tanks).

भजेच्छिदोंऽशैरथ तैर्विमिश्रै रूपं भजेत् स्यात् परिपूर्तिकालः ।। CIII ।।

One divided by the sum of the reciprocals (of the times taken by the sources to fill up a pool) is the time of filling (the pool when the sources are used simultaneously).

Comment: Suppose the sources take times $t_1, t_2, \ldots$ to fill up a reservoir. If they are simultaneously used then the time taken $= \dfrac{1}{\dfrac{1}{t_1} + \dfrac{1}{t_2} + \cdots}$. We use the same method at present. Of course at Bhāskarācārya's time there were no taps.

Example

ये निर्झरा दिनदिनार्धतृतीयषष्ठैः
संपूर्णयन्ति हि पृथग्पृथगेव मुक्ताः।
वापीं यदा युगपदेव सखे विमुक्ताः
ते केन वासरलवेन तदा वदाऽशु।। CIV ।।

Four streams flow into a pool and separately they take $1, \frac{1}{2}, \frac{1}{3}, \frac{1}{6}$ days respectively. If all the four are used simultaneously, find the time required to fill up the pool.

Comment: As explained in the previous stanza.

$$\text{Time} = \frac{1}{1+2+3+6} = \frac{1}{12}\text{th day.}$$

The four streams can fill up 1, 2, 3, 6 pools in one day and so together they can fill up 12 pools in one day. So time for one pool = $\frac{1}{12}$ day.

Formula concerning parts of grain

पण्यैः स्वमूल्यानि भजेत्स्वभागैः
हत्वा तदैक्येन भजेच्च तानि।
भागांश्च मिश्रेण धनेन हत्वा
मूल्यानि पण्यानि यथाक्रमं स्युः।। CV ।।

Form products of each cost, the reciprocal of the respective measure of grain (available for the cost) and the corresponding quantity (or proportion of the grain to be purchased). Each product (thus obtained) multiplied by the total price to be paid and divided by the sum of the products will be the price of each quantity of the (corresponding) grain to be purchased. (Sum of these quotients is naturally the total price to be paid.) Quantity (or weight) of a grain (when proportions of grains to be purchased are given) is the (proportion of the) quantity multiplied by the total price to be paid and divided by the sum of the products (obtained earlier).

Comment: For the ith grain, suppose, g_i measures cost d_i and we need quantity q_i. Then for money worth d units,

$$\text{cost of } i^{th} \text{ grain} = \frac{\frac{d_i}{g_i} \times g_i \times d}{\frac{d_1}{g_1} q_1 + \frac{d_2}{g_2} \times q_2 + \cdots}$$

Now to find the quantity of grain g_i in the mixture substitute the proportion of the grain g_i for $\frac{d_i}{g_i} q_i$ in the above formula.

Example 1

सार्धं तण्डुलमानकत्रयमहो द्रम्मेण मानाष्टकं
मुद्गानां च यदि त्रयोदशमिता एता वणिक्काकिणीः।
आदायाऽऽर्पय तण्डुलांशयुगलं मृद्गैकभागान्वितम्
क्षिप्रं क्षिप्रभुजो व्रजेमहि यतः सार्थोऽग्रतो यास्यति।। CVI ।।

$3\frac{1}{2}$ units of rice and 8 units of *mūng* beans can be bought for one *dramma.* O grocer, I have 13 *kākiṇīs* and I want rice and *mūng* beans (a kind of kidney beans) in the proportion 2:1. Quickly give me the grains so that we can cook, eat and start on our next journey by caravan.

Comment:

$$\text{Cost of rice} = \frac{\frac{1}{7/2} \times \frac{2}{1} \times \frac{13}{64}}{\frac{1}{7/2} \times \frac{2}{1} + \frac{1}{8} \times 1} = \left(\frac{4}{7} \times \frac{13}{64}\right) \div \left(\frac{4}{7} + \frac{1}{8}\right) = \frac{1}{6} \textit{ dramma}$$

$$\text{cost of } \textit{mūng} \text{ beans} = \left(\frac{1}{8} \times \frac{1}{1} \times \frac{13}{64}\right) \div \left(\frac{4}{7} + \frac{1}{8}\right) = \frac{7}{192} \textit{ dramma}$$

$$\text{quantity of rice} = \left(2 \times \frac{13}{64}\right) \div \left(\frac{4}{7} + \frac{1}{8}\right) = \frac{7}{12} \text{ measures}$$

$$\text{quantity of } \textit{mūng} \text{ beans} = \left(1 \times \frac{13}{64}\right) \div \left(\frac{4}{7} + \frac{1}{8}\right) = \frac{7}{24} \text{ measures.}$$

We can prove this formula by Algebra. Let x be the measure of *mūng* beans. Then rice = 2x measures.

Cost of *mūng* beans $= \frac{x}{8}$ *drammas*

cost of rice $= \frac{4x}{7}$ *drammas.*

So the total cost $= \frac{x}{8} + \frac{4x}{7} = \frac{39}{56}x$ *drammas* which is given as $\frac{13}{64}$ *dramma.*

So $\frac{39\text{ x}}{56} = \frac{13}{64} \quad \therefore\ \text{x} = \frac{7}{24}$ measures.

Example 2

कर्पूरस्य वरस्य निष्कयुगलेनैकं पलं प्राप्यते
वैश्यानंदनचन्दनस्य च पलं द्रम्माष्टभागेन चेत्।
अष्टांशेन तथाऽगरोः पलदलं निष्केण मे देहि तान्
भागैरेककषोडशाष्टकमितैर्धूपं चिकीर्षाम्यहम्।। CVII ।।

Good quality camphor costs 2 N (*niṣkas*) for 1 P (*pala*), sandalewood costs $\frac{1}{8}$ D (*drammas*) for 1 P and aloe costs $\frac{1}{8}$ D for $\frac{1}{2}$ P. O you grocer's son (who pleases his mother) what quantity of each is available in the proportion 1:16:8 for 1 N?

Comment: Noting that 1 N = 16 D we get

$$\text{cost of camphor} = \frac{\frac{32}{1}\times 1}{\frac{32\times 1}{1}+\frac{1}{8}\times 16+\frac{1}{8}\times 8\times 2}\times 16$$

$$= \frac{32\times 16}{36} = \frac{128}{9} = 14\frac{2}{9}\text{ D}.$$

$$\text{cost of sandalewood} = \frac{\frac{1}{8}\times 16\times 16}{36} = \frac{8}{9}\text{ D}$$

$$\text{cost of aloe} = \frac{\frac{1}{8}\times 8\times 2\times 16}{36} = \frac{8}{9}\text{ D}$$

$$\text{quantity of camphor} = 14\frac{2}{9}\div 32 = \frac{4}{9}\text{ P}$$

$$\text{quantity of sandalewood} = \frac{8}{9}\div\frac{1}{8} = \frac{64}{9}\text{ P}$$

$$\text{quantity of aloe} = \frac{8}{9}\div\frac{1}{4} = \frac{32}{9}\text{ P}.$$

We can work out this problem by algebra. Suppose camphor is x P. Then sandalewood = 16x P, aloe 8x P.

Total cost = $32x + 16x \times \frac{1}{8} + 8x \frac{1}{4} = 36x = 16$ D

$\therefore x = \frac{4}{9}$ P.

Formula for Exchange of Jewels

नरघ्नदानोनितरत्नशेषैरिष्टे हृते स्युः खलु मूल्यसंख्याः।
शेषैर्हृते शेषवधे पृथक्स्थैरभिन्नमूल्यान्यथवा भवन्ति।। CVIII ।।

(Suppose n persons possess $a_1, a_2, \ldots a_n$ jewels respectively. Suppose each one gives q jewels to each of the others and then their jewels are worth the same.) From the numbers of jewels, subtract the product of number of persons and the number of jewels donated (uniformly) to each of them. An assumed number (which could be the L.C.M. of the remainders or any multiple of the L.C.M.) divided by the remainders will give the prices of the (corresponding) jewels.

Comment: Compute $a_i - nq$; the price of a jewel with the ith person is $\frac{x}{a_i - nq}$ where x is the supposed worth of each. It can also be found by $\frac{\prod_{i=1}^{n}(a_i - nq)}{a_i - nq}$

The price of the jewels will naturally depend upon the supposed number x. Perhaps the answers will be fractions. If the L.C.M. of the remainders $\{(ai - nq) : 1 \leqq i \leqq n\}$ is divided by $a_i - nq$, the prices will be integral.

Example (Exchange of Jewels)

माणिक्याष्टकमिन्द्रनीलदशकं मुक्ताफलानां शतम्
सद्वज्राणि च पञ्च रत्नवणिजामेषां चतुर्णां धनम्।
संगस्नेहवशेन ते निजधनाद्दत्त्वैकमेकं मिथो
जातास्तुल्यधनाः पृथग्वद सखे तद्रत्नमूल्यानि मे।। CIX ।।

Four merchants had 8 rubies, 10 sapphires, 100 pearls and 5 diamonds respectively. They became friends on their journey and each one gave to each of the others one jewel from his lot. This made their jewels worth the same. O friend! find out the prices of these jewels.

Comment: We give three methods.

(1) *Bhāskarācārya's:* Jewels 8, 10, 100, 5. Number of gifts q=1, number of persons n = 4. So the remainders are $8 - 1 \times 4 = 4$, $10 - 4 \times 1 = 6$, $100 - 4 \times 1 = 96$, $5 - 4 \times 1 = 1$. If we suppose x = 96, the prices of jewels are ruby = $\frac{96}{4} = 24$, sapphire = $\frac{96}{6} = 16$, diamond = $\frac{96}{1} = 96$. Here 96 is the L.C.M. of the remainders. If we follow the second method we get the product of the remainders = $4 \times 6 \times 96 \times 1 = 2304$ and we get the prices of the jewels in the proportion 576, 384, 24, 2304 and if we choose 1 as the price of a pearl, the proportion is 24, 16, 1, 96 as before.

(2) After the transfer of the jewels, the 1st merchant had 5 R and one each of the other three or 4 R and one each of the four type of jewels. 2nd merchant had 6 S and one each of the four type of jewels. 3rd merchant had 96 P and one each of the four type of jewels. 4th merchant had 1 D and one each of the four type of jewels. Since their worth is the same, we get 4 R = 6 S = 96 P = 1 D. We choose the price of 1 pearl to be 1 Rupee and we get the answer as before.

(3) *Algebraic:* If the prices of the jewels are a, b, c, d respectively, the equations are

$5a + b + c + d = a + 7b + c + d = a + b + 97c + d = a + b + c + 2d$

$\therefore 4a = 6b = 96c = d$. If we choose c = 1 Rs., we get the same answer as before.

Formula to find the Weight of Pure Gold

सुवर्णवर्णाहतियोगराशौ स्वर्णैक्यभक्ते कनकैक्यवर्णः।
वर्णो भवेच्छोधितहेमभक्ते वर्णोद्धृते शोधितहेमसंख्या।। CX ।।

(To find the weight and fineness of gold if two or more kinds of gold are melted together.) The fineness of the mixture equals the sum of the products of weight and fineness of the constituents, divided by the total weight. If the same sum is divided by the weight of pure gold, one gets the fineness of pure gold and if it is divided by the fineness, the result will be the weight of pure gold.

Comment: If the fineness of the mixture is x and weight y then xy = $\sum_{i=1}^{n} x_i y_i$ and the result follows.

Note that $y = y_1 + ... + y_n$.

Example

विश्वार्करुद्रदशवर्णसुवर्णमाषाः दिग्वेदलोचनयुगप्रमिताः क्रमेण।
आवर्तितेषु वद तेषु सुवर्णवर्णः तूर्णं सुवर्णगणितज्ञ वणिग्भवेत् कः ।। CXI ।।

O golden mathematician, four types of gold 10 M (*māṣās*) of 13 C (carats), 4 M of 12 C, 2 M of 11 C and 4 M of 10 C are melted together to form a new one. Find its fineness. If this is purified and 16 M gold is obtained, what is its fineness? If the mixed gold when purified has 16 C fineness, what is its weight?

Comment: Sum of the products
$= 13 \times 10 + 12 \times 4 + 11 \times 2 + 10 \times 4$
$= 130 + 48 + 22 + 40 = 240.$

$$\text{Fineness of the mixture} = \frac{240}{10+4+2+4} = \frac{240}{20} = 12 \text{ C}$$

which naturally lies between 10 C and 13 C.

$$\text{Fineness of the purified gold} = \frac{240}{16} = 15 \text{ C.}$$

$$\text{If the fineness is 16 C, weight} = \frac{240}{16} = 15 \text{ M.}$$

Here 5 M of impurities were burnt out.

Formula: From Mixture to Component

स्वर्णैक्यनिघ्नाद्युतिजातवर्णात् सुवर्णतद्वर्णवधैक्यहीनात्।
अज्ञातवर्णाग्निजसंख्ययाप्तम् अज्ञातवर्णस्य भवेत् प्रमाणम् ।। CXII ।।

Multiply the weight of the mixture and its fineness. Subtract from this, the sum of the products of fineness and the corresponding weights of the given components. The remainder divided by the weight of the component (whose fineness is not known) gives the desired fineness.

Comment: (Using the notation of the previous stanza):

$$x_j = \frac{xy - \sum_{i \neq j} x_i y_i}{y_j}.$$

Example

दशेशवर्णा वसुनेत्रमाषा अज्ञातवर्णस्य षडेतदैक्ये।
जातं सखे द्वादशकं सुवर्णम् अज्ञातवर्णस्य वद प्रमाणम्।। CXIII ।।

8 M (*māṣās*) of 10 C (carats), 2 M of 11 C, 6 M of unknown C are mixed together to form 16 M of 12 C.
What is the fineness of the unknown?

Comment: Here $x = \frac{16 \times 12 - 8 \times 10 - 2 \times 11}{6}$

$$= \frac{192 - 80 - 22}{6} = 15\,C.$$

स्वर्णैक्यनिघ्नो युतिजातवर्णः स्वर्णघ्नवर्णैक्यवियोजितोऽसौ।
अहेमवर्णाग्निजयोगवर्णविश्लेषभक्तोऽविदिताग्निजं स्यात्।। CXIV ।।

Multiply the fineness of the mixtures by the sum of the known weights of the components of the mixture. Obtain the difference of this and the sum of the products of fineness and the corresponding weights of the components. The outcome divided by the difference of fineness of the component (whose weight is not known) and that of the mixture gives the desired weight.

[In the above notation, $y_3 = \frac{x(y_1 + y_2) - x_1 y_1 - x_2 y_2}{x_3 - x}$]

Comment: $x(y_1 + y_2 + y_3) = x_1y_1 + x_2y_2 + x_3y_3$.

Example

दशेन्द्रवर्णा गुणचन्द्रमाषाः किञ्चित्तथा षोडशकस्य तेषाम्।
जातं युतौ द्वादशकं सुवर्णं कतीह ते षोडशवर्णमाषाः।। CXV ।।

3 M (*māṣās*) of 10 C (carats), 1 M of 14 C are mixed with some quantity of 16 C. If the fineness of the mixture is 12 C, find the weight of 16 C gold.

Comment: By Bhāskarācārya's method

$$y_3 = \frac{12\,(3+1) - 10\times 3 - 14\times 1}{16-12} = \frac{48-44}{4} = 1\,\text{M}.$$

Weights of Components

साध्येनोनोऽनल्पवर्णो विधेयः साध्यो वर्णः स्वल्पवर्णोनितश्च।
इष्टक्षुण्णे शेषके स्वर्णमाने स्यातां स्वल्पानल्पयोर्वर्णयोस्ते।। CXVI ।।

(If fineness of a mixture and its two components are given, to find weights (indeed the ratio of the weights) of the components.) Obtain: the difference of the fineness of the mixture and that of the component with greater fineness; and the difference of the fineness of the mixture and that of the component with smaller fineness. The two differences multiplied by an assumed number will give weights of the components.

[Suppose $x_1 > x_2$. $\dfrac{y_1}{y_2} = \dfrac{x - x_2}{x_1 - x}$.

$$y_1 = \left(\frac{x - x_2}{x_1 - x_2}\right)(y_1 + y_2).]$$

Comment: $x(y_1 + y_2) = x_1y_1 + x_2y_2$.

Example

हाटकगुटिके षोडशदशवर्णे तद्युतौ सखे जातम्।
द्वादशवर्णसुवर्णं ब्रूहि तयोः स्वर्णमाने मे।। CXVII ।।

O friend, two small balls of gold of 16 C and 10 C were melted together to form a mixture of 12 C. Find out the weights of the balls.

Comment: If the mixture weighs y M (*māṣās*), then

$$y_1 = y \times \frac{12-10}{16-10} = \frac{y}{3},\; y_2 = y \times \frac{16-12}{16-10} = \frac{2y}{3}$$

So the proportion is 1:2.

CHAPTER 26

Combinations

एकाद्येकोत्तरा अंका व्यस्ता भाज्याः क्रमस्थितैः।
परः पूर्वेण संगुण्यस्तत्परस्तेन तेन च ।। CXVIII।।
एकद्वित्र्यादिभेदाः स्युरिदं साधारणं स्मृतम्।
छन्दश्चित्त्युत्तरे छन्दस्युपयोगोऽस्य तद्विदाम् ।। CXIX।।
मूषावहनभेदादौ खण्डमेरौ च शिल्पके।
वैद्यके रसभेदीये तन्नोक्तं विस्तृतेर्भयात् ।। CXX।।

Starting with the number n write down n, (n – 1), (n – 2), ... Divide them by 1, 2, 3, . . . to get $\frac{n}{1}, \frac{n-1}{2}, \frac{n-2}{3}, \ldots$. Then the number of combinations of n things taken 1, 2, 3, . . . at a time are $\frac{n}{1}, \frac{n(n-1)}{1 \times 2}, \frac{n(n-1)(n-2)}{1 \times 2 \times 3}, \ldots$ respectively. Or the number of combinations of n things taken r at a time are [n (n – 1) (n – 2) . . . (n – r + 1)] ÷ [1 × 2 × 3 × . . . × r]. This can be used to solve the problem when r = 1, 2, 3, ...

This is useful in prosody to discover all possible meters, in architecture, medical sciences, *Khanḍmeru*[*], chemical compositions etc. I am omitting these (applications) for the sake of brevity.

Comment: Modern notation is $\binom{n}{r} = \dfrac{n(n-1)\ldots(n-r+1)}{r!}$ The first line of the stanza is not clear, as no mention is made regarding the starting number in the product.

Example 1

प्रस्तारे मित्र गायत्र्याः स्युः पादे व्यक्तयः कति।
एकादिगुरवश्चाशु कथ्यतां तत्पृथक् पृथक्।। CXXI ।।

O friend, there are six letters in the fourth line of *gāyatrī* meter. If we choose only 1 g (*'guru'* means a long vowel in Prosody), 2 g or 3 g, how many meters are possible.

Comment: If the fourth line contains six ℓ (*'laghu'* ≡ short vowel), then the only combination is $\ell\ell\ell\ell\ell\ell$. If we have one g and five ℓ, then (i) $g\ell\ell\ell\ell\ell$ (ii) $\ell g\ell\ell\ell\ell$ (iii) $\ell\ell g\ell\ell\ell$ (iv) $\ell\ell\ell g\ell\ell$ (v) $\ell\ell\ell\ell g\ell$ (vi) $\ell\ell\ell\ell\ell g$ are possible. If there are 2 g and 4 ℓ, combinations will be 15.

As seen above if there is 1 g, number of meters = $\binom{6}{1} = 6$.

$$\text{For } 2\text{ g}, \quad \binom{6}{2} = \frac{6\times 5}{1\times 2} = 15$$

$$3\text{ g}, \quad \binom{6}{3} = \frac{6\times 5\times 4}{1\times 2\times 3} = 20$$

$$4\text{ g}, \quad \binom{6}{4} = \binom{6}{2} = 15$$

$$5\text{ g}, \quad \binom{6}{5} = \binom{6}{1} = 6$$

$$6\text{ g}, \quad \binom{6}{6} = 1.$$

---

[*] *Khanḍmeru* is Pascal's triangle.

Also $\binom{n}{0}+\binom{n}{1}+\cdots+\binom{n}{n}=2^n$

If we take the combinations of all the four lines as above, the total number of combinations will be $64 \times 64 \times 64 \times 64 = 16777216$.

Example 2

एकद्वित्र्यादिमूषावहनमितिमहो ब्रूहि मे भूमिभर्तुः
हर्म्ये रम्येऽष्टमूषे चतुरविरचितेश्लक्ष्णशालाविशाले।
एकद्वित्र्यादियुक्त्या मधुरकटुकषायाम्लकक्षारतिक्तैः
एकस्मिन् षड्रसैः स्युर्गणक कति वद व्यंजने व्यक्तिभेदाः ।। CXXII ।।

A king had a beautiful palace with eight doors. Skilled engineers had constructed four open squares which were highly polished and huge. In order to get fresh air, 1 door, 2 doors, 3 doors, . . . are opened, how many different types of breeze arrangements are possible?

How many kinds of relishes can be made by using 1, 2, 3, 4, 5 or 6 types from a sweet, bitter, astringent, sour, salty, hot substances?

Comment: As per the above formula, the possibilities are $\binom{8}{0},\binom{8}{1},\binom{8}{2},\ldots,\binom{8}{8}$. These are 1, 8, 28, 56, 70, 56, 28, 8, 1 respectively.

Total number $= 2^8 = 256$.

The number of different relishes (*chutneys*) $= 2 + 12 + 30 + 20 = 64 = 2^6$.

Pascal's Triangle

Bhāskarācārya has referred to *"khanḍameru"*, which is known as Pascal's triangle. This was known to ancient Indian mathematicians. *Piṅgalācārya* had used it in the construction of meters (in Prosody). In this triangle, every line is connected to the next one. Each line consists of the binomial coefficients as in Newton's binomial theorem. See the following scheme.

Pascal's triangle

$(x + y)^0$	0						
$(x + y)^1$	1	1					
$(x + y)^2$	1	2	1				
$(x + y)^3$	1	3	3	1			
$(x + y)^4$	1	4	6	4	1		
$(x + y)^5$	1	5	10	10	5	1	
$(x + y)^6$	1	6	15	20	15	6	1

For example, $(x + y)^4 = x^4 + 4x^3y + 6x^2y^2 + 4xy^3 + y^4$.

Also note 5 = 1 + 4, 10 = 4 + 6, 10 = 6 + 4, 5 = 4 + 1 in the 6th line.

This is based on the formula $\binom{n}{r} + \binom{n}{r-1} = \binom{n+1}{r}$. In Bhāskarā-cārya's times, the above was written like a *'meru'* (mountain).

$$
\begin{array}{ccccccccccc}
 & & & & & 0 & & & & & \\
 & & & & 1 & & 1 & & & & \\
 & & & 1 & & 2 & & 1 & & & \\
 & & 1 & & 3 & & 3 & & 1 & & \\
 & 1 & & 4 & & 6 & & 4 & & 1 & \\
1 & & 5 & & 10 & & 10 & & 5 & & 1
\end{array}
$$

CHAPTER 27

Progressions (Series)

सैकपदघ्नपदार्धमथैकाद्यंकयुतिः किल संकलिताख्या।
सा द्वियुतेन पदेन विनिघ्नी स्यात् त्रिहृता खलु संकलितैक्यम् ।। CXXIII।।

$$1 + 2 + 3 + \ldots + n = \frac{n(n+1)}{2}.$$

$$1 + (1 + 2) + (1 + 2 + 3) + \ldots + (1 + 2 + \ldots + n)$$
$$= \frac{n(n+1)(n+2)}{6}.$$

Example

एकादीनां नवान्तानां पृथक् संकलितानि मे।
तेषां सङ्कलितैक्यानि प्रचक्ष्व गणक द्रुतम्।। CXXIV।।

Find $1 + 2 + \ldots + 9$
and $1 + (1 + 2) + \ldots + (1 + 2 + \ldots + 9)$.

Comment: As per (CXXIII) the answers are $\frac{9 \times 10}{2} = 45$ and $\frac{9 \times 10 \times 11}{6} = 165$ respectively.

In current notation we write $\sum_{i=1}^{n} i, \sum_{i=1}^{n} \frac{i(i+1)}{2}$.

The first one is an Arithmetic Progression, and is a special case of

$$a + (a + d) + (a + 2d) + \ldots + (a + (n - 1)d) = \frac{n}{2}(2a + (n - 1)d).$$

Formulae for Σn^2, Σn^3 can be established by Mathematical induction and are found in current books.

द्विघ्नपदं कुयुतं त्रिविभक्तं संकलितेन हतं कृतियोगः।
संकलितस्य कृतेः सममेकाद्यंकघनैक्यमुदाहृतमाद्यैः ।। CXXV ।।

$$1^2 + 2^2 + \ldots + n^2 = \frac{n(n+1)(2n+1)}{6}.$$

$$1^3 + 2^3 + \ldots + n^3 = \left[\frac{n(n+1)}{2}\right]^2.$$

Arithmetic Progression

व्येकपदघ्नचयो मुखयुक स्यादन्त्यधनम् मुखयुग्दलितं तम्।
मध्यधनं पदसंगुणितं तत्सर्वधनं गणितं च तदुक्तम्।। CXXVI ।।

If the first term is a and the common difference (C.D.) is d, the nth term of the A.P. is given by $\ell = a + (n - 1)d$. The sum of the n terms is given by $\frac{n}{2}(a + \ell)$. Here $\frac{a + \ell}{2}$ is the middle term.

Example 1

तेषामेव च वर्गैक्यं घनैक्यं च वद द्रुतम्।
कृतिसंकलनामार्गे कुशला यदि ते मतिः ।। CXXVII ।।

Tell me the sums of $1^2 + \ldots + 9^2$ and $1^3 + \ldots + 9^3$.

Comment: $1^2 + \ldots + 9^2 = \frac{9(10)(19)}{6} = 285$

and $1^3 + \ldots + 9^3 = \left[\frac{9 \times 10}{2}\right]^2 = 2025.$

Example 2

आदिः सप्त चयः पंच गच्छोऽष्टौ यत्र तत्र मे।
मध्यान्त्यधन-संख्ये के वद सर्वधनं च किम् ।। CXXVIII।।

If the first term of an A.P. is 7 the common difference is 5 and the number of terms is 8, find the middle term, the last term and the sum.

Comment: $\ell = a + (n - 1)d = 7 + 7 \times 5 = 42$ and middle term = $\frac{7+42}{2} = \frac{49}{2}$ which is not a term in the A.P.

Sum $= 8 \times \frac{49}{2} = 196.$

Example 3

आद्ये दिने द्रम्मचतुष्टयं यो दत्त्वा दिनेभ्योऽनुदिनं प्रवृतः।
दातुं सखे पंचचयेन पक्षे द्रम्मा वद द्राक्कति तेन दत्ताः ।। CXXIX।।

A gentleman gave 4 D (*drammas*) as charity (to a Brahmin) on the first day. For fifteen days he continued his charity, everyday increasing his contribution by 5 D over the previous day. Find out the total charity.

Comment: Here a = 4, C.D. = d = 5, n = 15.

Last term $\ell = 4 + 14(5) = 74.$

Middle term $m = \frac{a+\ell}{2} = \frac{4+74}{2} = 39.$

Total charity $= 15 \times 39 = 585$ D.

Formula to find the first term of an A.P.

गच्छहृते गणिते वदनं स्यात् व्येकपदघ्नचयार्धविहीने ।। CXXX।।

To find the first term of an A.P., divide the given sum by the number of terms. Multiply the number of terms minus one by half the common difference, and subtract this result from the quotient already obtained. [Thus $a = \frac{S}{n} - \frac{(n-1)\,d}{2}$.]

Example

पंचाधिकं शतं श्रेढीफलं सप्तपदं किल।
चयं त्रयं वयं विद्मो वदनं वद नंदन ।। CXXXI ।।

O son, find the mouth (the first term a) of an A.P. whose S = 105, n = 7, d = 3.

Comment: $a = \frac{105}{7} - \frac{6 \times 3}{2} 15 - 9 = 6$.

Formula to find C.D. in an A.P.

गच्छहृतं धनमादिविहीनं व्येकपदार्धहृतं च चयः स्यात्।। CXXXII ।।

Divide the sum by the number of terms (of an A.P.) and subtract the first term from this quotient. This result divided by half of the number of terms minus one is the common difference. [Thus $d = \frac{\frac{S}{n} - a}{(n-1)/2}$.]

Example

प्रथममगमदह्ना योजने यो जनेशः
तदनु ननु कयाऽसौ ब्रूहि यातोऽध्ववृद्ध्या।
अरिकरिहरणार्थं योजनानामशीत्या
रिपुनगरमवाप्तः सप्तरात्रेण धीमन्।। CXXXIII ।।

To capture enemy elephants, a king covers 2 Y (*yojanas*) on the first day and then increases his distance by A.P. on subsequent days. If he travels 80 Y in 7 days, O intelligent boy, find out the extra distance each day.

Comment:

$$d = \frac{\frac{80}{7} - 2}{\frac{7-1}{2}} = \frac{66}{7} \times \frac{2}{6} = \frac{22}{7} = 3\frac{1}{7} \text{ Y} .$$

Formula to find the number of terms of an A.P.

श्रेढीफलादुत्तरलोचनघ्नात् चयार्धवक्त्रान्तरवर्गयुक्तात्।
मूलं मुखोनं चयखंडयुक्तम् चयोद्धृतं गच्छमुदाहरन्ति।। CXXXIV।।

Add the square of the difference between the first term and half the C.D. to the sum (of the given A.P.) multiplied by twice the C.D. Then find (the positive) square root of this result. Subtract the first term from and add half the C.D. to this square root. The resultant divided by the C.D. is number of terms (in the A.P.). [Thus

$$n = \frac{\sqrt{2Sd + (a - \frac{d}{2})^2} - (a - \frac{d}{2})}{d}.]$$

Comment: $S = \frac{n}{2}\,[2a + (n-1)d]$

$$\therefore dn^2 + n\,(2a - d) - 2S = 0.$$

Bhāskarācārya's formula is the positive root of the above quadratic equation. He does not admit a negative d. Also n should be obviously positive.

Example

द्रम्मत्रयं यः प्रथमेह्नि दत्त्वा दातुं प्रवृत्तो द्विचयेन तेन।
शतत्रयं षष्ट्यधिकं द्विजेभ्यो दत्तं कियद्भिर्दिवसैर्वदाऽऽशु।। CXXXV।।

A donor gave 3 D (*drammas*) in charity to a Brahmin on the first day. He continued increasing his donation each day by 2 D. If the total amount paid by him equals 360 D, how many days did he give in charity?

Comment:

$$n = \frac{\sqrt{2 \times 360 \times 2 + \left(3 - \frac{2}{2}\right)^2} - \left(3 - \frac{2}{2}\right)}{2}$$

$$= \frac{\sqrt{1440 + 4} - 2}{2} = \frac{38 - 2}{2} = 18.$$

Geometric Progression (G.P.)

विषमे गच्छे व्येके गुणकः स्थाप्यः समेऽर्धिते वर्गः ।
गच्छक्षयांतमंत्यात् व्यस्तं गुणवर्गजं लं यत्तत् ।
व्येकं व्येकगुणोद्धृतमादिगुणं स्यात् गुणोत्तरे गणितम् ।। CXXXVI ।।

If n, the number of terms in a G.P. is odd, then (n – 1) is called a 'multiplier' (M) and if it is even, $\frac{n}{2}$ is called a 'square' (S) [Bhāskarācārya's terminology]. Now beginning with n, continue this process (n – 1) for odd and $\frac{n}{2}$ for even until 0 is reached. Then keep the common ratio (C.R.) r agianst 0 and start writing M or S in the opposite direction. Carry out the operations and then subtract 1 from the final result and divide it by (r – 1). The result is the required sum.

Comment: Suppose a = 1, n = 31, r = 2. Then following the above method, we get the table:

31	30	15	14	7	6	3	2	1	0
	M	S	M	S	M	S	M	S	M

Beginning with 31 we write 31 – 1 = 30, $\frac{30}{2}$ = 15, 15 – 1 = 14, $\frac{14}{2}$ = 7, . . . until we reach 0. Then in the second line we begin with M and alternately write M and S.

Now we perform the indicated operations:
M = multiplier, S = square as shown below.

			Index
30	M	2147483648	31
15	S	1073741824	30
14	M	32768	15
7	S	16384	14
6	M	128	7
3	S	64	6
2	M	8	3
1	S	4	2
0	M	2	1

To reach the 31st power, here Bhāskarācārya has given a shorter method by squaring at even numbers and multiplying by r at odd ones. In practice, multiplication might be easier than squaring.

As is given in standard texts:

$$S = a + ar + ar^2 + \ldots + ar^{n-1}$$

$$rS = ar + ar^2 + \ldots + ar^{n-1} + ar^n$$

$$S(r-1) = ar^n - a \qquad \therefore S = \frac{a(r^n - 1)}{r-1}.$$

Example 1

पूर्वं वराटकयुगं येन द्विगुणोत्तरं प्रतिज्ञातम्।
प्रत्यहमर्थिजनाय स मासे निष्कान् ददाति कति।। CXXXVII।।

A gentleman gave 2 C (cowries) in charity the first day and thereafter he gave everyday twice of what he gave the previous day. How many N (*niṣkas*) did he give away in one month?

Comment: Here a = 2, r = 2, n = 30.

$$S = \frac{2(2^{30} - 1)}{2-1} = 2(1073741824 - 1) = 2147483646 \text{ C}$$

= 104857 N, 9 *drammas,* 9 *paṇas,*
2 *kākiṇīs,* 6 C.

Example 2

आदिर्द्वयं सखे वृद्धिः प्रत्यहं त्रिगुणोत्तरा।
गच्छ सप्तदिनं यत्र गणितं तत्र किं वद।। CXXXVIII।।

If a = 2, r = 3, n = 7, O friend, what is the sum?

Comment: $S = \frac{2(3^7 - 1)}{3-1} = \frac{2(2187-1)}{3-1} = 2186$.

Formula for the number of metres (in Prosody)

पादाक्षरमितगच्छे गुणवर्गफलं चये द्विगुणे।
समवृत्तानां संख्या तद्वर्गो वर्गवर्गश्च।।
स्वस्वपदोनौ स्यातां अर्धसमानां च विषमाणाम्।। CXXXIX।।

[In every metre there are four quarters. If the order of short and long vowels is the same in each of the four quarters, the metre is known as 'even' (*samavṛtta*).] The following formula gives the number of even metres. Take as n the number of letters in each quarter and r = 2 (for there are only two kinds of vowels, short and long). The total number of possible metres = 2^n.

[If the first and the third quarters are similar and so are the second and the fourth, it is called 'semi-even' (*ardhasamavṛtta*).] In this case the total number of possible metres = $2^{2n} - 2^n$.

[If the number of letters is the same but all the four quarters are different it is called 'odd metre' (*viṣamavṛtta*).] In this case the total number = $(2^n)^4 - (2^n)^2$.

Comment: Suppose there are n letters in each line. Each letter can be short or long and so there are two choices. According to the fundamental theorem there are 2^n possibilities in the even metre.

There are $2^n \times 2^n$ possibilities in the 1st two (as well as 3rd and 4th). From these we subtract the even case 2^n to get the number of semi-even cases: $(2^n)^2 - 2^n = 2^n (2^n - 1)$.

In the odd metre case, there are $2^n \times 2^n \times 2^n \times 2^n$ possibilities from which we subtract the first two cases: $2^n + 2^n (2^n - 1)$. So the number $= (2^n)^4 - [2^n + 2^{2n} - 2^n] = (2^n)^4 - 2^{2n}$.

Example

समानामर्धतुल्यानां विषमाणां पृथक् पृथक्।
वृत्तानां वद मे संख्यामनुष्टुप्छंदसि द्रुतम्।। CXL।।

Find the number of even, semi-even and odd metres in *'anuṣṭupa'* metre.

Comment: There are 8 letters. So

number of even metres = $2^8 = 256$

number of semi-even metres = $(2^8)^2 - 2^8 = 65280$

number of odd metres = $(2^8)^4 - (2^8)^2 = 4294901760$.

CHAPTER 28

Mensuration

Measurement of Sides and Areas

इष्टो बाहुर्यः स्यात् तत्स्पर्धिन्यां दिशीतरो बाहुः ।
त्र्यस्त्रे चतुरस्त्रे वा सा कोटिः कीर्तिता तज्ज्ञैः ।। CXLI ।।

In a right-angled triangle, one of the sides is called the base (*bhuja* or *bāhu*) and the side perpendicular to it is called the altitude (*koṭi*). The same terminology holds for a rectangle.

Comment:

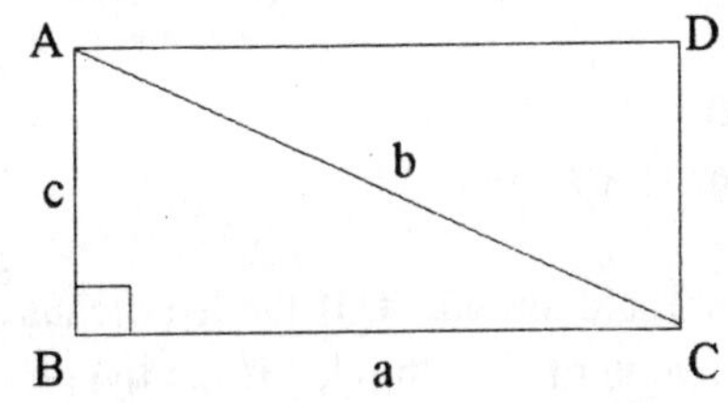

BC is the base.

AB is the altitude.

AC is the hypotenuse.

तत्कृत्योर्योगपदं कर्णो दोःकर्णवर्गयोर्विवरात्।
मूलं कोटिः कोटिश्रुतिकृत्योरंतरात् पदं बाहुः ।। CXLII ।।

In a right-angled triangle, square root of the sum of squares of the base and the altitude is the hypotenuse, and square root of the difference between squares of the hypotenuse and base (respectively altitude) is the altitude (resp. base). [Thus $b = \sqrt{c^2 + a^2}$, $c = \sqrt{b^2 - a^2}$, $a = \sqrt{b^2 - c^2}$.]

Comment: This is known as Pythagoras' Theorem. This was known in India since the time of *Śulvasūtrakāras* (3000-800 B.C.) whereas Pythagoras published it in 560 B.C. Of course, no proofs were given at Bhāskarācārya's times. Proofs of this theorem are found in textbooks.

राश्योरन्तरवर्गेण द्विघ्ने घाते युते तयोः।
वर्गयोगो भवेदेवं तयोर्योगान्तराहतिः।
वर्गान्तरं भवेदेवं ज्ञेयं सर्वत्र धीमाता ।। CXLIII ।।

If twice the product of two numbers is added to the square of their difference, the result is the sum of their squares. So also if the sum of two numbers is multiplied by their difference, the result is the difference of their squares. [Thus $a^2 + b^2 = (a - b)^2 + 2ab$ and $a^2 - b^2 = (a + b)(a - b)$.]

Example 1

कोटिश्चतुष्टयं यत्र दोस्त्रयं तत्र का श्रुतिः।
कोटिं दोष्कर्णतः कोटिश्रुतिभ्यां च भुजं वद।। CXLIV ।।

Find the hypotenuse if the base is 3 and altitude 4. If the hypotenuse and the base are 5, 3 what is the length of the altitude? If the hypotenuse and the altitude are 5, 4, what is the base?

Comment: $3^2 + 4^2 = 5^2$ $\therefore$ the hypotenuse = 5.

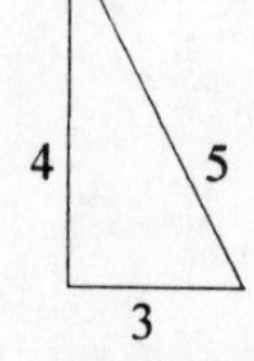

Example 2

सांघ्रित्रयमितो बाहुर्यत्र कोटिश्च तावती।
तत्र कर्णप्रमाणं किं गणक ब्रूहि मे द्रुतम् ।। CXLV।।

If in a right-angled triangle, the base is $3\frac{1}{4}$ and the altitude $3\frac{1}{4}$ find the hypotenuse.

Comment: Hypotenuse = $\sqrt{\frac{169}{16}+\frac{169}{16}}=\frac{13\sqrt{2}}{4}$.

वर्गेण महतेष्टेन हताच्छेदांशयोर्वधात्।
पदं गुणपदक्षुण्णच्छिद्भक्तं निकटं भवेत् ।। CXLVI।।

[In this stanza Bhāskarācārya explains a method of finding an approximate square root of a number which is not a perfect square.]

To compute (approximately) $\sqrt{\frac{a}{b}}$ choose a large square number x. Then compute approximately $\sqrt{abx}$ and divide by $b\sqrt{x}$.

Comment:

$$\sqrt{\frac{a}{b}} = \frac{\sqrt{abx}}{b\sqrt{x}} .$$

In the above Example 2, $\sqrt{\frac{169}{8}}=\frac{13\sqrt{2}}{4} \doteq \frac{13\,(1\cdot414)}{4}=\frac{13\,(141\cdot4)}{400}$

$$= \left\{4\frac{476}{800}+\frac{1}{200}\right\} \text{ approx.}$$

By Bhāskarācārya's method, choose $x = 10000 = 100^2$ and

$$\sqrt{\frac{169}{8}}=\sqrt{\frac{169\times8\times10000}{8\times8\times10000}} \doteq \frac{3677}{800}=4\frac{477}{800}.$$

Here we note that $\left.\begin{array}{l} 3677^2 = 13520329 \\ \text{and } 169\times8\times10000 = 13520000 \end{array}\right\}$.

Bhāskarācārya was aware of the fact that $\sqrt{2}$, $\sqrt{3}$, $\sqrt{17}$,... are irrational numbers and so gave a method to compute their values approximately. In his times, 'decimal fractions' were not known. They were introduced in Italy for the first time in the sixteenth century. Until then only vulgar fractions were made use of in all mathematical calculations. Bhāskarācārya rationalizes the denominator.

Formula to find hypotenuse and altitude of a right-angled triangle given its base

इष्टेन भुजोऽस्माद् द्विगुणेननिघ्नादिष्टस्य कृत्यैकवियुक्तयाप्तम्।
कोटिः पृथक् सेष्टगुणाभुजोना कर्णो भवेत् त्र्यस्त्रमिदं तुजात्यम्।। CXLVII।।

इष्टो भुजस्तत्कृतिरिष्टभक्ता द्विः स्थापितेष्टोनयुताऽधिता वा।
तौ कोटिकर्णौ इति कोटितो वा बाहुश्रुती चाकरणीगते स्तः।। CXLVIII।।

a = base, b = hypotenuse, c = altitude (see fig. of stanza CXLI). Choose any convenient number y. Then given a to find b, c; $c = \frac{2ay}{y^2-1}$, $b = \frac{a(y^2+1)}{y^2-1}$.

OR: $b = \frac{1}{2}\left(\frac{a^2}{y}+y\right)$ and $c = \frac{1}{2}\left(\frac{a^2}{y}-y\right)$.

Comment: (1) Here $b^2 = \frac{a^2(y^2+1)^2}{(y^2-1)^2}$ and

$$a^2+c^2 = a^2+\frac{4a^2y^2}{(y^2-1)^2} = \frac{a^2(y^2-1)^2+4a^2y^2}{(y^2-1)^2} = \frac{a^2(y^2+1)^2}{(y^2-1)^2} = b^2.$$

(2) In the second case,

$$a^2+c^2 = a^2+\frac{1}{4}\left(\frac{a^2}{y}-y\right)^2 = \frac{4a^2y^2+a^4-2a^2y^2+y^4}{4y^2} = \frac{(a^2+y^2)^2}{4y^2} = b^2.$$

Example

भुजे द्वादशके यौ यौ कोटिकर्णौ अनेकधा।
प्रकाराभ्यां वद क्षिप्रं तौ तावकरणीगतौ ।। CXLIX।।

If the base of a right triangle is 12, find its altitude and hypotenuse in integers.

Comment: *1st Method.* Choose y so that $y^2 - 1$ divides 2ay. Here a = 12 and we choose y = 2.

$$c = \frac{2ay}{y^2 - 1} = \frac{2 \times 12 \times 2}{4 - 1} = 16, b = \frac{a(y^2 + 1)}{y^2 - 1} = \frac{12(5)}{3} = 20 .$$

2nd Method. Here y must divide a^2 and we choose y = 6.

$$b = \frac{1}{2}\left(\frac{12^2}{6} + 6\right) = 15, c = \frac{1}{2}\left(\frac{12^2}{6} - 6\right) = 9.$$

If we take y = 2 we get b = 37, c = 35.

NOTE: These are known as (Pythagorean) triplets. Several solutions are possible: (i) (9, 12, 15), (ii) (12, 16, 20), (iii) (5, 12, 13), (iv) (12, 35, 37). These are the only possible ones. The first two are derived from (3, 4, 5) by multiplying by 3 and 4 respectively and so are called *secondary.* (3, 4, 5), (5, 12, 13), (12, 35, 37) are called *primary.* Such distinction was not made at Bhāskarācārya's times. There are well-known formulae to find the triplets:

$$[(2n + 1), (2n^2 + 2n), (2n^2 + 2n + 1)]$$

and $$[2n, n^2 - 1, n^2 + 1]; \left[n, \frac{n^2 - 1}{2}, \frac{n^2 + 1}{2}\right].$$

Formula to find two sides when the hypotenuse is given

इष्टेन निघ्नाद् द्विगुणाच्च कर्णादिष्टस्य कृत्येकयुजा यदाप्तम्।
कोटिर्भवेत् सा पृथगिष्टनिघ्नी तत्कर्णयोरन्तरमत्र बाहुः ।। CL ।।

Given hypotenuse b, choose a convenient number y. Then $a = \frac{2by}{y^2 + 1}$,

$$c = \frac{b(1 - y^2)}{(y^2 + 1)} .$$

Comment: $a^2 + c^2 = \frac{b^2 [4y^2 + 1 - 2y^2 + y^4]}{(y^2 + 1)^2} = b^2 .$

Example

पंचाशीतिमिते कर्णे यौ यावकरणीगतौ।
स्यातां कोटिभुजौ तौ तौ वद कोविद सत्वरम्।। CLI।।

If the hypotenuse is 85 find the other sides in integers.

Comment: Here we choose y = 2, since $y^2 + 1$ divides b.

$a = \frac{2 \times 85 \times 2}{5} = 68, c = \frac{85 \times 3}{5} = 51$. We could have chosen y = 4 and got a = 40, c = 75. By using (2n + 1), $(2n^2 + 2n)$, $(2n^2 + 2n + 1)$ we find $2n^2 + 2n + 1 = 85 \therefore n = 6$ and we get (13, 84, 85).

Another formula when hypotenuse is given

इष्टवर्गेण सैकेन द्विगुणः कर्णोऽथवा हृतः।
फलोनः श्रवणः कोटिः फलमिष्टगुणं भुजः।। CLII।।

$$a = b - \frac{2b}{y^2 + 1}, c = \frac{2by}{y^2 + 1}.$$

Comment:

$$a = \frac{b(y^2 - 1)}{y^2 + 1}, c = \frac{2by}{y^2 + 1}$$

$$a^2 + c^2 = b^2.$$

These are the same (except for a sign) as in (CL), but a and c are interchanged.

Formula to construct right triangles with integral sides

इष्टयोराहतिर्द्विघ्नी कोटिर्वर्गान्तरं भुजः।
कृतियोगस्तयोरेवं कर्णश्चाकरणीगतः।। CLIII।।

Given any two numbers x, y, there is a right trinagle with a = 2xy, $c = x^2 - y^2$ (x > y) and $b = x^2 + y^2$.

Comment: $b^2 = (x^2 + y^2)^2 = (x^2 - y^2)^2 + 4x^2y^2 = c^2 + a^2$.
This formula can be used whether or not x, y are integers.

Example

यैयैस्त्र्यस्त्रं भवेज्जात्यं कोटिदोश्श्रवणैः सखे।
त्रीनप्यविदितानेतान् क्षिप्रं ब्रूहि विचक्षण।। CLIV।।

If all the three sides of a right triangle are unknown, O friend, tell (me) quickly their values.

Comment: Take any two unequal numbers, say $x = 2$, $y = 1$. Then $a = 4$, $c = 3$, $b = 5$. If $x = 3$, $y = 2$, we get $a = 12$, $c = 5$, $b = 13$. One can construct many such triplets.

वंशाग्रमूलान्तरभूमिवर्गो वंशोद्धृतस्तेन पृथग्युतोनौ।
वंशो तदर्धे भवतः क्रमेण वंशस्य खंडे श्रुतिकोटिरूपे।। CLV।।

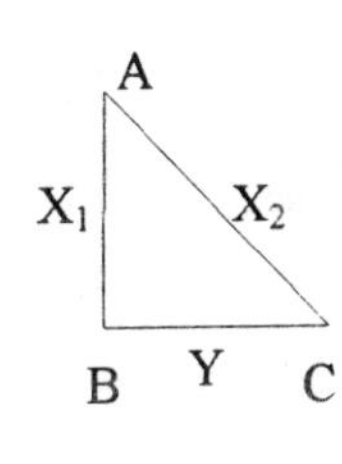

Suppose a bamboo BAC is broken at A and the part AC touches the ground at C. Given the length x of the bamboo and the distance BC = y, find x_1 and x_2 by: $x_1 = \frac{1}{2}(x - \frac{y^2}{x})$,

$$x_2 = \frac{1}{2}(x + \frac{y^2}{x}).$$

Comment: Here $x_1 + x_2 = x$ and $x_1^2 + y^2 = x_2^2$. So

$$x_1^2 = x_2^2 - y^2 = (x - x_1)^2 - y^2 = x^2 - 2xx_1 + x_1^2 - y_2^2$$

$$\therefore 2xx_1 = x^2 - y^2 \qquad \therefore x = \left(\frac{x^2 - y^2}{2x} \right).$$

$$\text{Then } x_2 = x - x_1 = \frac{x^2 + y^2}{2x}.$$

Example 1

यदि समभुवि वेणुर्द्वित्रिपाणिप्रमाणो
गणक पवनवेगादेकदेशे स भग्नः।
भुवि नृपमितहस्तेष्वंग लग्नं तदग्रम्
कथय कतिषु मूलादेष भग्नः करेषु।। CLVI।।

A bamboo 32 C (cubits) high, standing on a level ground, was broken by strong winds. The tip of the bamboo touched the ground 16 C from the foot of the bamboo. Then, O mathematician, tell (me) the height of the point where the bamboo broke.

Comment: Here x = 32, y = 16.

$$x_1 = \frac{1}{2}\left(32 - \frac{16^2}{32}\right) = \frac{1}{2}(32 - 8) = 12.$$

Example 2

स्तंभस्य वर्गोऽहिबिलान्तरेण भक्तः फलं व्यालबिलान्तरालात् ।
शोध्यं तदर्धप्रमितः करैः स्यात् बिलाग्रतो व्यालकलापियोगः ।। CLVII ।।

[Near the foot of a pole, there is a snake burrow. A peacock jumps from the top of the pole and pounces on the snake. To find the distance of the snake from the pole.]
Given AB = x, AC = y, to find AD = r and BD = DC = s.

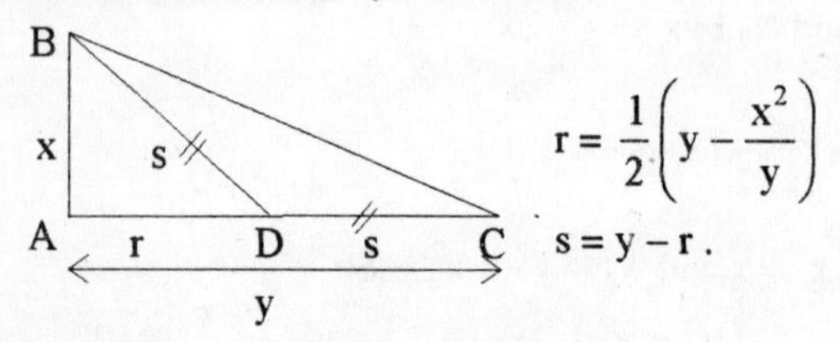

Comment: Here $r^2 = s^2 - x^2 = (y - r)^2 - x^2$

$$\therefore r^2 = y^2 - 2yr + r^2 - x^2$$

$$\therefore r = \frac{1}{2}(y - \frac{x^2}{y}).$$

Example

अस्ति स्तंभतले बिलं तदुपरि क्रीडाशिखंडी स्थितः
स्तंभे हस्तनवोच्छ्रिते त्रिगुणितस्तंभप्रमाणान्तरे ।
दृष्ट्वाहिं बिलमाव्रजन्तमपतत् तिर्यक् स तस्योपरि ।
क्षिप्रं ब्रूहि तयोर्बिलात्कतिमितैः साम्येन गत्योर्युतिः ।। CLVIII ।।

There was a snake burrow at the foot of a pole, on the top of which was seated a domesticated peacock. The pole was 9 C (cubits) high. The peacock saw a snake crawl towards the post at a distance of 27 C from the post. The peacock pounced on the snake at the same speed as the snake's crawl and caught it at a certain distance from the pole. Find the distance quickly.

Comment: Here y = 27, x = 9

$$r = \frac{1}{2}\left(27 - \frac{9^2}{27}\right) = \frac{1}{2}(27 - 3) = 12\text{ C}.$$

The peacock and the snake travelled equal distances of 15 C.

NOTE: Some modern mathematicians may object to the above solution on the grounds that the path of the peacock is a parabola due to the force of gravity. However it is wrong to conclude that Bhāskarācārya was ignorant of gravity as the following verse (Golādhyāya – Chapter *Bhuvanakoṣa,* Stanaza 6) shows.

आकृष्टशक्तिश्च मही तया यत् खस्थं गुरुं स्वाभिमुखं स्वशक्त्या।
आकृष्यते तत्पततीव भाति समे समन्तात् क्व पतत्वियं खे।।

That is the earth attracts inert bodies in space towards itself. The attracted body appears to fall down on the earth. Since the space is homogeneous, where will the earth fall?

Also the peacock does not fall down like an inert coconut falling down from a palm. He purposefully pounces on the snake and keeps his speed equal to that of the snake. Such conditions arise in many mathematical problems. So we cannot say that the problem is wrong. Also since the problem is on right triangles, we are justified in taking BD to be a straight line.

Formula to find height of lotus and depth of a pond

भुजाद्वर्गितात्कोटिकर्णान्तराप्तं द्विधा कोटिकर्णान्तरेणोनयुक्तम्।
तदर्धे क्रमात् कोटिकर्णौ भवेतामिदं धीमता वेत्य सर्वत्र योज्यम्।। CLIX।।

[Height of lotus flower is AC = AD = x. When the lotus moves by wind and its top touches the water at D, BD = r. BC = y.]

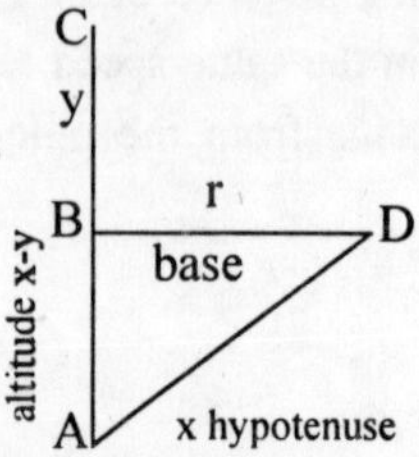

C
y
r
B D
base
altitude x-y
A x hypotenuse

Depth of water $AB = \frac{1}{2}\left(\frac{r^2}{y} - y\right)$.

Height of the lotus

$$AC = x = \frac{1}{2}\left(\frac{r^2}{y} + y\right).$$

सखे पद्मतन्मज्जनस्थानमध्यं भुजः कोटिकर्णान्तरं पद्म दृश्यम्।
नलः कोटिरेतन्मितं स्याद् यतोऽम्भो वदैवं समानीय पानीयमानम्।। CLX।।

O friend, the distance between B, the point where the lotus stem meets the water surface and D where the tip touches the water level, is called the base. The height of the lotus above the water surface is AD – AB = hypotenuse – altitude. Altitude is the depth of water.

Comment: Here $(x - y)^2 + r^2 = x^2$

$$\therefore x^2 - 2xy + y^2 + r^2 = x^2$$

$$\therefore x = \frac{1}{2}\left(\frac{r^2}{y} + y\right)$$

$$AB = x - y = \frac{1}{2}\left(\frac{r^2}{y} + y\right).$$

Example

चक्रक्रौंचाकुलितसलिले क्वापि दृष्टं तडागे
तोयादूर्ध्वं कमलकलिकाग्रं वितस्तिप्रमाणम्।

मन्दं मन्दं चलितमनिलेनाऽऽहतं हस्तयुग्मे
तस्मिन्मग्नं गणक कथय क्षिप्रमम्बुप्रमाणम्।। CLXI।।

In a lake there were a large number of ruddy geese and cranes. A lotus whose height above the water surface was one *vīta* ($\frac{1}{2}$ C (Cubit)), and its tip, bent by a rustling wind, sank at a distance of 2 C. O mathematician! tell (me) quickly the depth of water.

Comment: Here $y = \frac{1}{2}, r = 2$.

$$\text{Depth of water} = \frac{1}{2}\left(\frac{4}{1/2} - \frac{1}{2}\right) = \frac{15}{4}\text{C}.$$

$$\text{Length of lotus stem} = \frac{1}{2}\left(\frac{4}{1/2} + \frac{1}{2}\right) = \frac{17}{4}\text{C}.$$

द्विनिघ्नतालोच्छ्रितिसंयुतं यत्सरोऽन्तरं तेन विभाजितायाः।
तालोच्छ्रितेस्तालसरोन्तरघ्न्या उड्डीयमानं खलु लभ्यते तत्।। CLXII।।

[A monkey jumps from a point M on a palm tree AM to D and then onto a well. B. To find MD.]

$$r = \frac{cx}{2c + x}.$$

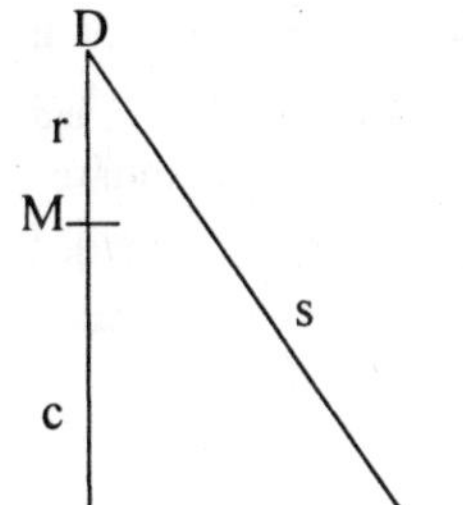

Comment: One monkey travels MD + DB = r + s and the other MA + AB = c + x and r + s = c + x . Also

$s^2 = x^2 + (c + r)^2$

$s^2 = x^2 + c^2 + r^2 + 2cr$ and

$r^2 + s^2 + 2rs = c^2 + x^2 + 2cx$

$\therefore 2cx - 2cr - r^2 = r^2 + 2rs$

$\therefore 2cx - 2cr = 2r^2 + 2r(c + x - r)$

$= 2r^2 + 2rc + 2rx - 2r^2$

$\therefore 2cx = 4cr + 2rx$

$\therefore r = \frac{cx}{2c + x}$ = Height of Jump.

Example

वृक्षाद्धस्तशतोच्छ्रयाच्छतयुगे वापीं कपिः कोऽप्यगात्
उत्तीर्याथ परो द्रुतं श्रुतिपथात्प्रोड्डीय किंचित द्रुमात्।
जातैवं समता तयोर्यदि गतावुड्डीयमानं कियत्
विद्वंश्चेत्सुपरिश्रमोऽस्ति गणिते क्षिप्रं तदाचक्ष्व मे।। CLXIII।।

There was a palm tree 100 C (cubits) high and there was a well at a distance of 200 C from the tree. Two monkeys were on the top of the tree. One of them came down the tree and walked to the well. The other one jumped up and then pounced on the well along the hypotenuse. If both covered equal distances, find the length of the second monkey's jump.

Comment: Jump $= \dfrac{100 \times 200}{2 \times 100 + 200} = 50$ C.

An objection may be raised that the second monkey traverses a parabolic path. But as per the conditions on his movement, the path is to be taken a straight line.

Formula: To solve a right triangle given its hypotenuse (b) and the sum of the other two (a + c)

कर्णस्य वर्गाद् द्विगुणाद्विशोध्यो दोःकोटियोगः स्वगुणोऽस्य मूलम्।
योगो द्विधा मूलविहीनयुक्तः स्याता तदर्धे भुजकोटिमाने।। CLXIV।।

[To find the base and altitude when their sum and hypotenuse of a right triangle are given.] Subtract, from twice the square of the hypotenuse, the square of the sum of the base and altitude. The square-root of this result subtracted from (resp. added to) the sum of the sides and divided by two happens to be the length of the base (resp. altitude). [Thus given b and a + c = s,

$$c = \frac{s + \sqrt{2b^2 - s^2}}{2} \text{ and } a = \frac{s - \sqrt{2b^2 - s^2}}{2}.]$$

Note: If the hypotenuse and the difference of two sides are given then, to find out the base (resp. altitude), this difference should be subtracted from (resp. added to) the square-root of {twice the square of the hypotenuse minus the square of the given difference} and divided by two; (see also the comment of stanza CLXVI).

Comment: $b^2 = a^2 + c^2 = (a + c)^2 - 2ac = s^2 - 2ac$ and

$$(c - a)^2 = (a + c)^2 - 4ac = s^2 + 2(b^2 - s^2) = 2b^2 - s^2$$

$$\therefore c - a = \sqrt{2b^2 - s^2} \text{ and}$$

$$c + a = s$$

Solving these equations, we have the formulae.

Bhāskarācārya calls the smaller a the base and larger c the altitude. This problem can also be solved by quadratic equations.

Example

दश सप्ताधिकाः कर्णस्त्र्यधिका विंशतिः सखे।
भुजकोटियुतिर्यत्र तत्र ते मे पृथक् वद।। CLXV।।

If in a right triangle, the hypotenuse is 17 and the sum of the other two sides is 23, find the base and the altitude.

Comment:
$$c = \frac{23 + \sqrt{2 \times 17^2 - 23^2}}{2} = 15$$

and
$$a = \frac{23 - \sqrt{2 \times 17^2 - 23^2}}{2} = 8.$$

Example

दोःकोट्योरन्तरं शैलाः कर्णो यत्र त्रयोदश।
भुजकोटी पृथक्तत्र वदाऽऽशु गणकोत्तम।। CLXVI।।

If in a right triangle, the hypotenuse is 13 and the difference between the other two sides is 7, find the sides.

Comment: Here if $c - a = s = 7$ and $b = 13$,
$$c = \frac{s + \sqrt{2b^2 - s^2}}{2} = \frac{7 + \sqrt{2 \times 13^2 - 7^2}}{2} = 12$$

and
$$a = \frac{\sqrt{2b^2 - s^2} - s}{2} = \frac{17 - 7}{2} = 5.$$

NOTE: Khanapur Shastri used the same formula as in (CLXV) and got $a = -5$. We use a different formula. Although Bhāskarācārya knew about negative square root, he says that a base of right triangle cannot be negative and this is accepted by modern mathematicians.

अन्योन्यमूलाग्रगसूत्रयोगात्
वेण्योर्वधे योगहृते च लंबः।
वंशौ स्वयोगेन हृतौ अभीष्ट-
भूघ्नौ च लंबोभयतः कुखंडे।। CLXVII।।

[There are two vertical poles AB, CD. Strings AD and BC meet at E. To find EF = y given AB = r and CD = s.]

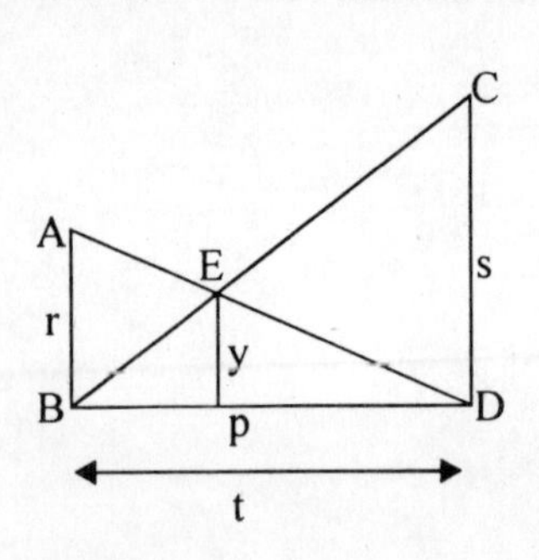

The product of heights of two poles divided by their sum is the perpendicular (EF). The products of the height of a pole (AB, say) and the distance between the poles is the distance (BF) between the pole and the perpendicular. [Thus

$$y = \frac{sr}{s+r},$$

$$BF = \frac{rt}{s+r}, \quad FD = \frac{st}{s+r}.]$$

Comment: $\Delta BEF, \Delta BCD$ are similar $\therefore \frac{y}{s} = \frac{BF}{t}$.

ΔEFD and ΔABD are similar $\therefore \frac{y}{r} = \frac{FD}{t}$.

Adding $\frac{y}{s} + \frac{y}{r} = \frac{BF+FD}{t} = 1 \therefore y = \frac{sr}{s+r}$. Another proof can be given by co-ordinate geometry. Clearly t is not needed to compute y.

Example

पंचदश-दशकरोच्छ्रायवेण्वोरज्ञातमध्यभूमिकयोः।
इतरेतरमूलाग्रगसूत्रयुतेर्लंबमानमाचक्ष्व।। CLXVIII।।

Two poles have heights 15 C (cubits) and 10 C respectively. Top of each is connected to the bottom of the other by two strings. Find the height of the point of intersection of the strings.

Comment: $y = \frac{sr}{s+r} = \frac{15 \times 10}{15+10} = 6\text{ C}$.

Existence of Triangles

धृष्टोद्दिष्टमृजुभुजं क्षेत्रं यत्रैकबाहुतः स्वल्पा।
तदितरभुजयुतिरथवा तुल्या ज्ञेयं तदक्षेत्रम्।। CLXIX।।

In a triangle (or a polygon) it is impossible for one side to be greater than the sum of the other sides. It is daring for anyone to say that such a thing is possible.

Comment: This is well-known result in Euclidean Geometry. It is unfortunate that ancient Indian mathematicians did not perceive the need to offer proofs for their assertions.

Example

चतुरस्त्रे द्विषट्त्र्यर्का भुजास्त्र्यस्त्रे त्रिषण्णव।
उद्दिष्टा यत्र धृष्टेन तदक्षेत्रं विनिर्दिशेत्।। CLXX।।

If an idiot says that there is a quadrilateral of sides 2, 6, 3, 12 or a triangle with sides 3, 6, 9, explain to him that they don't exist.

त्रिभुजे भुजयोर्योगेस्तदन्तरगुणो भुवा हृतो लब्ध्या।
द्विःस्था भूरूनयुता दलिताऽऽबाधे तयोः स्याताम्।। CLXXI।।

स्वाबाधाभुजकृत्योरन्तरमूलं प्रजायते लंबः।
लंबगुणं भूम्यर्धं स्पष्टं त्रिभुजे फलं भवति।। CLXXII।।

[In ΔABC, $AD \perp BC$. To find BD, DC.] In a triangle, (assuming a side as base) form the product of the sum and difference of (the remaning) two sides. Divide this product by the base. Add to and subtract from the base, this quotient. Divide the results thus obtained by two, and get the projections of the two sides (on the base).

Square-root of the difference of the squares of a side and its projection on the base is the length of the perpendicular (drawn from the vertex on the base). The product of half the base and the perpendicular is the area of the triangle. [Thus if $c > b$,

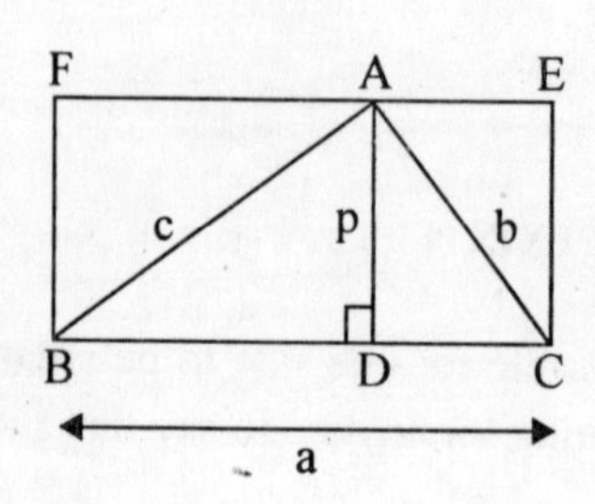

$$BD = \frac{1}{2}\left[a + \frac{(c+b)(c-b)}{a}\right]$$

$$DC = \frac{1}{2}\left[a - \frac{(c+b)(c-b)}{a}\right]$$

$$p^2 = c^2 - BD^2 = b^2 - DC^2.$$

Also Area $\Delta ABC = \frac{1}{2}$ pa.]

Comment: These are given in school geometry. If C is obtuse, some modifications are necessary. See Example 2 below.

Example 1

क्षेत्रे मही मनुमिता त्रिभुजे भुजौ तु
यत्र त्रयोदशतिथिप्रमितौ च मित्र।
तत्रावलंबकमितिं कथयावबाधे
क्षिप्रं तथा च समकोष्ठमितिं फलाख्याम्।। CLXXIII ।।

If the base of a triangle is 14 and the other sides are 13, 15, O friend, find the length of the perpendicular on the base, the lengths of two parts of the base into which it is divided by the foot of the perpendicular and the area of the triangle.

Comment: $BD = \frac{1}{2}\left[14 + \frac{(28)(2)}{14}\right] = 9,$

$$DC = \frac{1}{2}\left[14 - \frac{(28)(2)}{14}\right] = 5,$$

$$P = \sqrt{15^2 - 9^2} = 12$$

and $\Delta ABC = \frac{1}{2} \times 14 \times 12 = 84$ sq. units.

Example 2

दशसप्तदशप्रमौ भुजौ त्रिभुजे यत्र नवप्रमा मही।
अबधे वद लंबकं तथा गणितं गाणितिकाशु तत्र मे।। CLXXIV।।

A triangle has sides 10, 17 and base 9. O mathematician, find its altitude, the two projections (on the base) and the area.

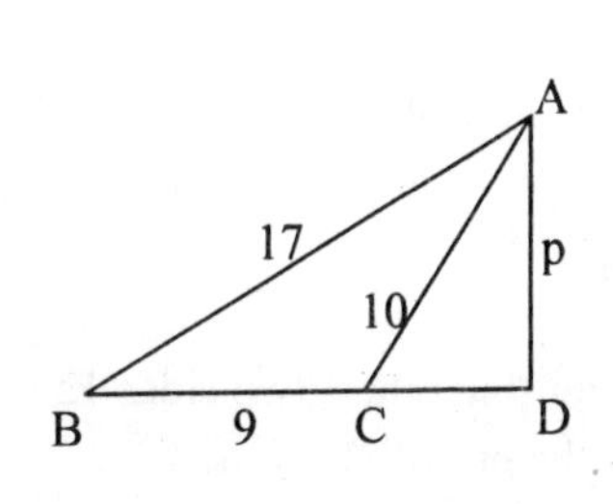

Comment: Here projection of

$$AB = BD = \frac{1}{2}\left[9 + \frac{27 \times 7}{9}\right] = 15,$$

projection of AC = CD

$$= \frac{1}{2}\left[\frac{27 \times 7}{9} - 9\right] = 6,$$

$$p = \sqrt{10^2 - 6^2} = 8$$

and area = ½ × 9 × 8 = 36 sq. units.

Here $A\hat{C}B$ is obtuse and so there is an appropriate modification.

Formulae: Areas of (cyclic) quadrilateral and triangle

सर्वदोर्युतिदलं चतुःस्थितं बाहुभिर्विरहितं च तद्वधात्।
मूलमस्फुटफलं चतुर्भुजे स्पष्टमेवमुदितं त्रिबाहुके।। CLXXV।।

From half the sum of all the (four) sides subtract each side separately and take their product. Square-root of this product is imprecise area of quadrilateral. (However) the precise area of a triangle ascends in this manner. [Let a, b, c, d be the sides of the quadrilateral and s = $\frac{a+b+c+d}{2}$ semi-perimeter. Imprecise area of a quadrilateral is

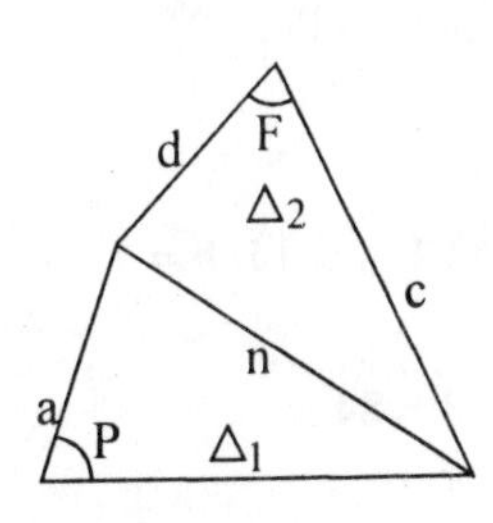

$= \sqrt{(s-a)(s-b)(s-c)(s-d)}$. Area of a triangle with sides a, b, c is obtained by putting d = 0, viz. $\sqrt{s(s-a)(s-b)(s-c)}$.]

Comment: The formula for the area of a quadrilateral is exact for a *cyclic* quadrilateral. Formula for the area of a triangle was discovered by Heron (200 B.C.). The former, as also pointed out by Bhāskarācārya, is not correct for a non-cyclic quadrilateral for which the correct formula is

$$\sqrt{(s-a)(s-b)(s-c)(s-d) - abcd\cos^2\left(\frac{P+F}{2}\right)}.$$

This derivation can be found in a standard trigonometry book.

Example 1

भूमिश्चतुर्दशकरा मुखमंकसंख्यं
बाहू त्रयोदशदिवाकरसंमितौ च।
लंबोऽपि यत्र रविसंख्यक एव तत्र
क्षेत्रे फलं कथय तत्कथितं यदाद्यैः।। CLXXVI ।।

A quadrilateral has base 14, upper side 9 and (the other) two sides 12 and 13. If the altitude is 12, find its area by the primitive method.

Comment: Clearly the quadrilateral is a trapezium since one side and the altitude are equal. Also we can break the trapezium into a rectangle and a right triangle.

$$\text{So the area} = 12 \times 9 + \frac{1}{2} \times 12 \times 5$$

$$= 108 + 30 = 138.$$

Or using the primitive formula:

$$\text{Area} = \sqrt{(s-a)(s-b)(s-c)(s-d)}$$

$$= \sqrt{(24-12)(24-9)(24-13)(24-14)}$$

$$= \sqrt{19800} \doteq 140.7\ .$$

We get imprecise value since the quadrilateral is not cyclic.

Example 2

We now find area of the triangle of stanza CLXXIII: a = 14, b = 13, c = 15, s = 21.

$$\text{Area} = \sqrt{s(s-a)(s-b)(s-c)} = \sqrt{21\times7\times6\times8} = 84\ .$$

चतुर्भुजस्यानियतौ हि कर्णौ कथं ततोऽस्मिन्नियतं फलं स्यात्।
प्रसाधितौ तच्छ्रवणौ यदाद्यैः स्वकल्पितौ तावितरत्र न स्तः।। CLXXVII ।।

The (primitive) formula for the area of a quadrilateral is not accurate. This is so because the lenghts of its diagonals are indeterminate. So how can we get an accurate value? Ancient mathematicians had fixed some values for the diagonals but they are not valid in all cases.

तेष्वेव बाहुष्वपरौ च कर्णावनेकधा क्षेत्रफलं ततश्च।। CLXXVIII ।।

Even if all the four sides of quadrilaterals are equal (i.e., a rhombus), diagonals can be different and consequently we get different areas.

Comment: Nowdays high school students know that three independent elements of a triangle are needed to fix a triangle. Similarly, five independent elements are necessary to determine a quadrilateral. In stanza CLXXVI, besides the four sides we are given one angle to be a right angle.

लंबयोः कर्णयोर्वैकं अनिर्दिश्यापरः कथम्।
पृच्छत्यनियत्वेऽपि नियतं चापि तत्फलम्।। CLXXIX ।।

स पृच्छकः पिशाचो वा वक्ता वा नितरां ततः।
यो न वेत्ति चतुर्बाहुक्षेत्रस्यानियतां स्थितिम्।। CLXXX ।।

If a perpendicular or a diagonal of a quadrilateral is not given, then its area is indeterminate. So one who asks such a question is a devil and one who can answer it must be a greater one because he does not know that even if all the four sides are given, the area is indeterminate.

इष्टा श्रुतिस्तुल्यचतुर्भुजस्य कल्प्याथ तद्वर्गविवर्जिता या।
चतुर्गुणा बाहुकृतिस्तदीयं मूलं द्वितीयश्रवणप्रमाणम्।। CLXXXI ।।

अतुल्यकर्णाभिहतिद्विभक्ता फलं स्फुटं तुल्यचतुर्भुजे स्यात्।
समश्रुतौ तुल्यचतुर्भुजे च तथाऽऽयते तद्भुजकोटिघातः।। CLXXXII ।।

चतुर्भुजेऽन्यत्र समानलंबे लंबेन निघ्नं कुमुखैक्यखंडम्।। CLXXXIII ।।

Subtract the square of a diagonal of a rhombus from four times the square of its base, and the square-root of the remainder is the other diagonal. The precise area of a rhombus is half the product of its two diagonals. In a rhombus of equal diagonals and rectangle, the area is the product of the adjacent sides. The area of a trapezium is the product of half the sum of the parallel sides and the perpendicular (distance between them). [Rhombus of sides = a and diagonals d_1, d_2 :

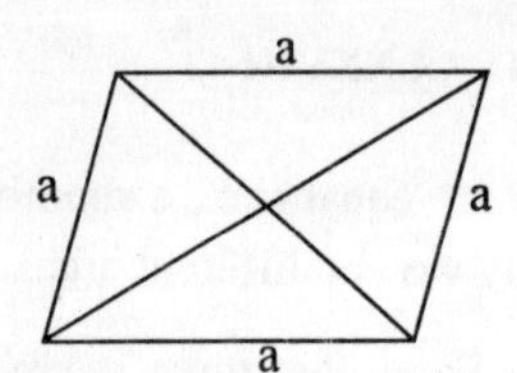

(Given a and d_1) $d_2 = \sqrt{4a^2 - d_1^2}$

(Given d_1, d_2) Area $= \frac{1}{2} d_1 d_2$.

Area of a rectangle = product of its adjacent sides.

Area of a trapezium $= \frac{1}{2}$ (sum of parallel sides) × height.]

Comment: These are standard formulae.

Example

क्षेत्रस्य पंचकृतितुल्यचतुर्भुजस्य
कर्णौ ततश्च गणितं गणक प्रचक्ष्व।
तुल्यश्रुतेश्च खलु तस्य तथाऽऽयतस्य
यद्विस्तृती रसमिताष्टमितं च दैर्घ्यम्।। CLXXXIV ।।

(1) O mathematician, in a rhombus of sides 25, find two diagonals and area.
(2) If the diagonals are equal, what is its area?
(3) Find the area of a rectangle if its base and vertical side are 6 and 8 respectively.

Comment: (1) Choose $d_1 = 30$. Here $a = 25$.

$d_2 = \sqrt{4 \times 25^2 - 30^2} = 40$

$\text{Area} = \frac{30 \times 40}{2} = 600$ sq. units.

If we choose $d_1 = 48$ then $d_2 = 14$ and area is 336 sq units. Thus a rhombus with given sides can have infinitely many pairs of diagonals and areas.

(2) Area = 25 × 25 = 625 sq. units.

(3) Area = 6 × 8 = 48 sq. units.

Example 1

क्षेत्रस्य यस्य वदनं मदनारितुल्यं
विश्वंभरा द्विगुणितेन मुखेन तुल्या।
बाहू त्रयोदशनखप्रमितौच लंबः
सूर्योन्मितश्च गणितं वद तत्र किं स्यात्।। CLXXXV ।।

The front of a field is eleven, its earth (base) is twice the front and the other two sides are thirteen and twenty. Twelve is the perpendicular (between the front and the base). Then find the area.

Comment: We are given a trapezium whose parallel sides are 11 and 22, other sides are 13, 20 and the altitude 12.

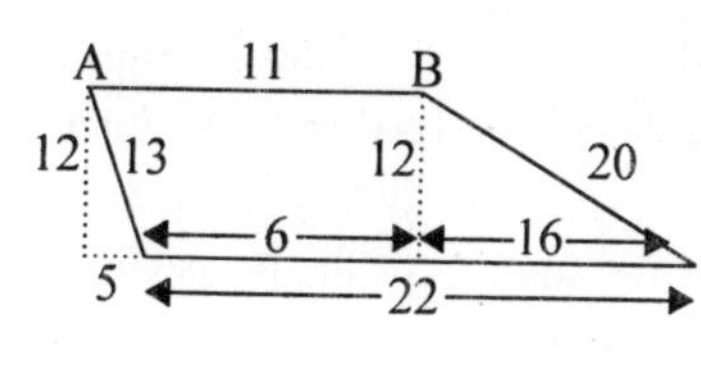

$$\text{Area} = \frac{1}{2}(11 + 22) \times 12 = 198 \text{ sq. units.}$$

[Here Bhāskarācārya has given one more datum than is necessary. Given the four sides we can compute the altitude, as is clear from the figure.]

Example 2

पंचाशदेकसहिता वदनं यदीयं
भूः पंचसप्ततिमिता प्रमितोऽष्टषष्ट्या।
सव्यो भुजो द्विगुणविंशतिसंमितोऽन्यः
तस्मिन्फलं श्रवणलंबमिती प्रचक्ष्व।। CLXXXVI ।।

In a quadrilateral, the upper side is 51, base 75, left side 68 and right side 40. Find its area, diagonals and altitude.

Comment: Here only 4 elements are given and so we suppose that diagonal AC = 77.

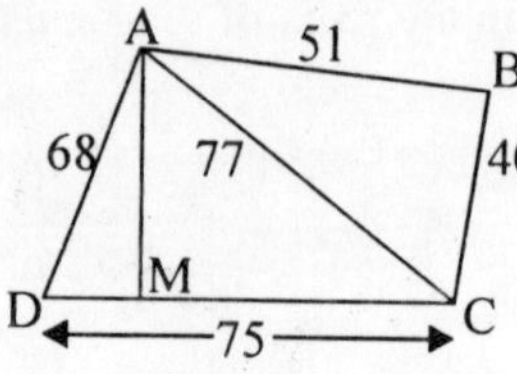

$$MC = \frac{1}{2}\left[\frac{(77+68)\,(77-68)}{75} + 75\right] = \frac{231}{5}$$

$$MD = 75 - MC = \frac{144}{5}$$

$$AM^2 = 68^2 - \frac{144^2}{25} \quad \therefore AM = \frac{308}{5}$$

Area of the quadrilateral = $\Delta ADC + \Delta ABC$

= 2310 + 924 = 3234 square units.

From the primitive formula $\sqrt{49 \times 77 \times 42 \times 66} = 3234$ which shows that the quadrilateral is cyclic.

If we take $C = 90°$, $BD^2 = 75^2 + 40^2 = 7225 \;\therefore BD = 85$.

Formula for a diagonal of a quadrilateral

यल्लंबलंबाश्रितबाहुवर्गविश्लेषमूलं कथिताऽबधा सा।
तदूनभूवर्गसमन्वितस्य यल्लंबवर्गस्य पदं स कर्णः।। CLXXXVII ।।

The projection of a side (on another side : = base of a quadrilateral) is square-root of the difference of the squares of the (given) perpendicular and the side. Diagonal (concerning with the two sides already considered) is square-root of the sum of the squares of the perpendicular and the difference of the base and projection.

Comment: (See figure of stanza CLXXXVI.)

$$DM = \sqrt{AD^2 - AM^2}\,,$$

$$CM = DC - DM$$

and

$$AC = \sqrt{MC^2 + AM^2}\,.$$

Formula to find the second diagonal

इष्टोऽत्र कर्णः प्रथमं प्रकल्प्यस्त्र्यस्रे तु कर्णोभयतः स्थिते ये।
कर्ण तयोः क्ष्मामितरौ च बाहू प्रकल्प्य लंबावबधे प्रसाध्ये।। CLXXXVIII ।।

आबाधयोरेककुप्स्थयोर्यत्स्यादन्तरं तत्कृतिसंयुतस्य।
लंबैक्यवर्गस्य पदं द्वितीयः कर्णो भवेत्सर्वचतुर्भुजेषु।। CLXXXIX ।।

[Sides of quadrilateral ABCD are given. Take some value for AC = 77.

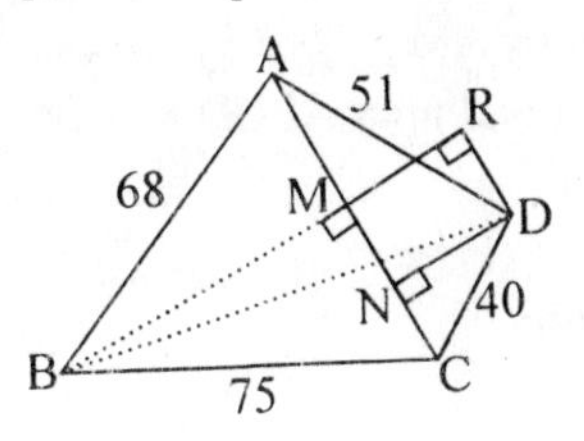

Sides of ΔABC and ΔADC are known. Suppose $BM \perp AC$, $DN \perp AC$. Compute AM and MC. Similarly altitudes can be computed. Thus we get AN, CN and DN in ΔACD. MN = DR.
BR = BM + MR = BM + DN.

Then $BD^2 = BR^2 + DR^2$.]
Assume some convenient value for one diagonal. If we take this diagonal as the base, we get two triangles. Then we get perpendicular BM as well as AM, MC, ND, AN, NC.
$BD^2 = BR^2 + RD^2 = [BM + ND]^2 + MN^2$.

Formula to find area of a quadrilateral

कर्णाश्रितंस्वल्पभुजैक्यमुर्वी प्रकल्प्य तच्छेषभुजौ च बाहू।
साध्योऽवलंबोऽथ तथान्यकर्णः स्वोर्व्याः कथंचिच्छ्रवणो न दीर्घः।। CXC।।

तदन्यलंबान्न लघुस्तथेदं ज्ञात्वेष्टकर्णः सुधिया प्रकल्प्यः।
त्र्यस्रे तु कर्णोभयतः स्थिते ये तयोः फलैक्यं फलमत्र नूनम्।। CXCI।।

Take the sum of the two smallest sides of a quadrilateral as the base and the remaining two sides to form a triangle. Draw the perpendicular to the original base (of the quadrilateral). [Although the length of the diagonal is variable or uncertain, it has maximum and minimum values.] The length of the diagonal is at most equal to the above base and at least equal to the length of the prependicular. So the length of the diagonal is choosen to lie between the two above values. A smart (student) can easily understand this. Taking this value for the diagonal, compute the sum of the area of the two triangles and this is the (approximate) value of the area of the quadrilateral.

Comment: ABCD is the given quadrilateral whose sides are given – 75, 40, 51, 68 but the fifth data is *not* given. Bhāskarācārya says that we can take the diagonal AC to have a length

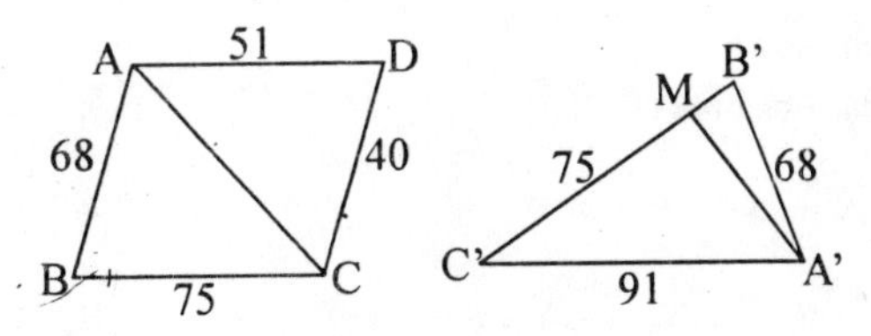

between the base 75 and the altitude A'M of $\Delta A'B'C'$ where $A'C' = 51 + 40 = 91$. Obviously $AC < 51 + 40 = 91$. Now $AM \doteq 66$. So we can choose $AC = 77$ and then compute $BD = 85$ and area $ABCD = 3234$ square units.

Formula to find the diagonals of a trapezium

समानलंबस्य चतुर्भुजस्य मुखोनभूमिं परिकल्प्य भूमिम्।
भुजौ भुजौ त्र्यस्रवदेव साध्ये तस्याबधे लंबमितिस्ततश्च।। CXCII ।।

आबाधयोना चतुरस्रभूमिः। तल्लंबवर्गैक्यपदं श्रुतिः स्यात्।।
समानलंबे लघुदोः कुयोगात्। मुखान्यदोः संयुतिरल्पिका स्यात्।। CXCIII ।।

Draw a triangle with the difference of parallel sides of a trapezium as its base and the slanting sides as other two sides. Then find lengths of the altitude and segments of the base (into which the altitude divides the base of the triangle). Next from the base of the trapezium subtract one of the segments. To the square of the remainder add the square of the altitude, and square-root of this sum is one of the diagonals. (It is clear that) a trapezium cannot exist unless the base added to the smaller side is together greater than the sum of the upper (parallel) side and the other side.

Comment: $AM \perp BC$, $AE \parallel DC$, $BF \perp DA$, $BE = BC - AD$.

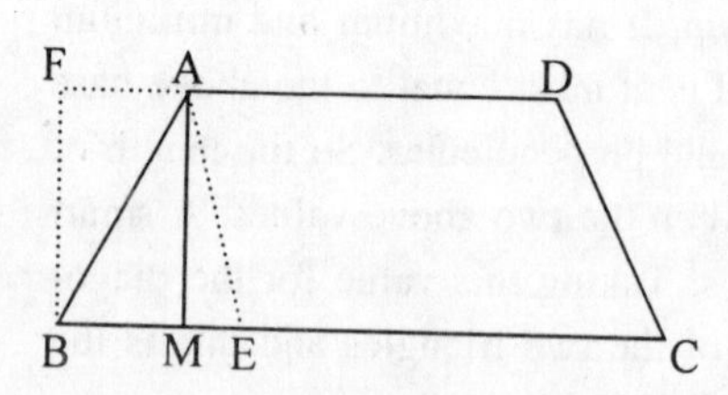

Compute AM, BM, ME in ΔABE.

$AC = \sqrt{AM^2 + MC^2}$

$BD = \sqrt{BF^2 + FD^2}$

$MC = ME + AD$

$FD = BM + AD$.

Example (Scalene Quadrilateral)

द्विपंचाशन्मितव्येकचत्वारिंशन्मितौ भुजौ।
मुखं तु पंचविंशत्या तुल्यं षष्ट्या मही किल।। CXCIV ।।

अतुल्यलंबकं क्षेत्रमिदं पूर्वैरुदाहृतम्।
षट्पंचाशत्रिषष्टिश्च नियते कर्णयोर्मिती।
कर्णौ तत्रापरौ ब्रूहि समलंबे च तच्छ्रुती।। CXCV।।

A quadrilateral has two opposite sides 39, 52, base 60 and upper side 25. Perpendiculars are unequal. Diagonals are 56 and 63. Taking these four sides but diagonals of different lengths, construct a quadrilateral. If this new figure is a trapezium, find the lengths of the diagonals. This example was given by earlier teachers.

Comment:

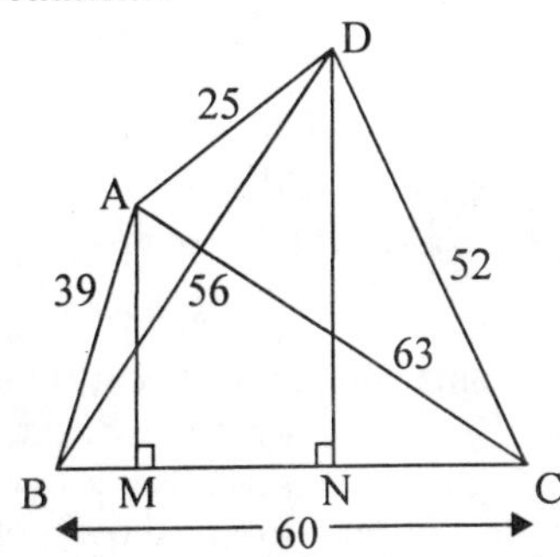

$$BM = \frac{1}{2}\left[60 - \frac{102 \times 24}{60}\right] = \frac{48}{5}$$

$$AM = \sqrt{AB^2 - BM^2} = \frac{189}{5}$$

$$BN = \frac{1}{2}\left[60 + \frac{108 \times 4}{60}\right] = \frac{168}{5}$$

$$DN = \sqrt{BD^2 - BN^2} = \frac{224}{5}$$

$$\therefore AM \neq DN$$

NOTE: If the reader knows Ptolemy's Theorem, it is easy to recognize ABCD as a cyclic quadrilateral.

$$AC \times BD = AD \times BC + AB \times DC$$

$$63 \times 56 = 25 \times 60 + 39 \times 52$$

i.e. $3528 = 1500 + 2028$ which is true.

In the above quadrilateral if BD = 32 (rather than 56), then what is AC?

$$BM = \frac{1}{2}\left[32 + \frac{64 \times 14}{32}\right] = 30 \text{ } (AM \perp BD)$$

$$AM^2 = 39^2 - 30^2 \quad \therefore AM \doteq 25.$$

If $CN \perp BD$, then

$$BN = \frac{1}{2}\left[32 + \frac{112 \times 8}{32}\right] = 30 \quad \therefore M = N.$$

That is, diagonals of ABCD intersect at right angles. Many commentators on *Līlāvatī* have missed this point. Now

$$MC^2 = 60^2 - 30^2 \qquad \therefore CM \doteq 52$$

$$AC \doteq 25 + 52 = 77.$$

Here the older writers should have taken approximate values rather than actual square roots.

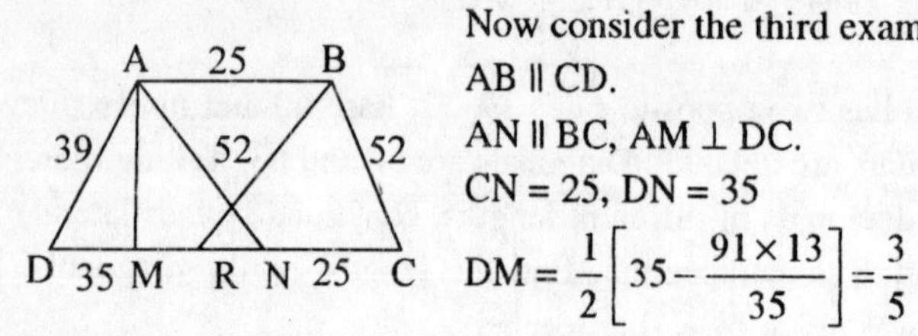

Now consider the third example wherein $AB \parallel CD$.

$AN \parallel BC$, $AM \perp DC$.

$CN = 25$, $DN = 35$

$$DM = \frac{1}{2}\left[35 - \frac{91 \times 13}{35}\right] = \frac{3}{5}$$

$$AM = \sqrt{39^2 - \frac{9}{25}} = \frac{1}{5}\sqrt{38016}$$

$$MC = 60 - \frac{3}{5} = \frac{297}{5}.$$

$$AC = \sqrt{AM^2 + MC^2} \doteq 71, \quad BD = \sqrt{\left(25 + \frac{3}{5}\right)^2 + AM^2} \doteq 46.7$$

$\text{Area} = \frac{1}{2}(60 + 25) \times \frac{195}{25} \doteq \frac{493}{2}$ square units, slightly different from the exact value.

Current methods of approximation differ from those of Bhāskarācārya.

Formula

The great Indian mathematician Brahmagupta has given this formula. He has shown that if a scalene quadrilateral is cyclic and its sides are given, then its diagonals can be calculated. It is a matter of pride for Indians that this formula was discovered in the sixth century whereas it took another thousand years for the west to rediscover this result. Bhāskarācārya has aptly praised Brahmagupta for this discovery.

कर्णाश्रितभुजघातैक्यमुभयथाऽन्योन्यभाजितं गुणयेत्।
योगेन भुजप्रतिभुजवधयोः कर्णौ पदे विषमे।। CXCVI ।।

(Diagonals of a cyclic quadrilateral). Find the sum of the products of sides holding a diagonal of a quadrilateral, and obtain a similar sum for the other diagonal. Multiply them by the number obtained on adding the products of the opposite sides. Divide each of the multiplied sums by the other (unmultiplied) sum. Square-roots of these quotients are diagonals.

Comment: Let m and n be the diagonals of a cyclic quadrilateral (see the adjoining figure). Then

$$n^2 = \frac{(ad + bc)(bd + ac)}{ab + cd}$$

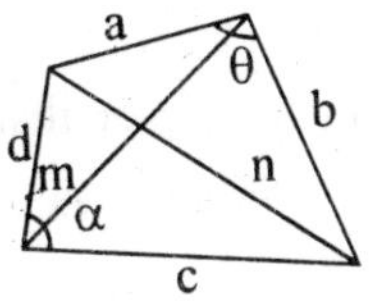

and

$$m^2 = \frac{(ab + cd)(bd + ac)}{ab + bc}$$

Since $\alpha + \theta = \pi$, $\cos\alpha + \cos\theta = 0$

i.e. $$\frac{c^2 + d^2 - n^2}{2cd} + \frac{a^2 + b^2 - n^2}{2ab} = 0$$

i.e. $ab(c^2 + d^2) + cd(a^2 + b^2) = n^2 (ab + cd)$.

Simplyfying, we get Brahmagupta's formula.

Bhāskarācārya's Method

अभीष्टजात्यद्वयबाहुकोटयः परस्परं कर्णहता भुजा इति।
चतुर्भुजं यद्विषमं प्रकल्पितं श्रुती तु तत्र त्रिभुजद्वयात्ततः।। CXCVII ।।

बाह्वोर्वधः कोटिवधेन युक्स्यादेका श्रुतिः कोटिभुजावधैक्यम्।
अन्या लघौ सत्यपि साधनेऽस्मिन् पूर्वैः कृतं यद्गुरु तन्न विद्मः।। CXCVIII ।।

(Here Bhāskarācārya gives his own simple method). Take two right triangles. Multiply the sides of one of the triangles by the hypotenuse of the other and the sides of the other triangle by the hypotenuse of the first. These are (the two sets of opposite) sides of a scalene quadrilateral. Further, sum of the product of the bases (shorter sides) of the triangles and the product of their altitudes is one diagonal (of the quadrilateral). The other diagonal is obtained by adding the product of one base and the other altitude to the product of one altitude and the other base. Thus the diagonals are easily obtained (by taking two right triangles). We do not know why (our) foremathematicians followed longer methods (for obtaining the diagonals of a quadrilateral).

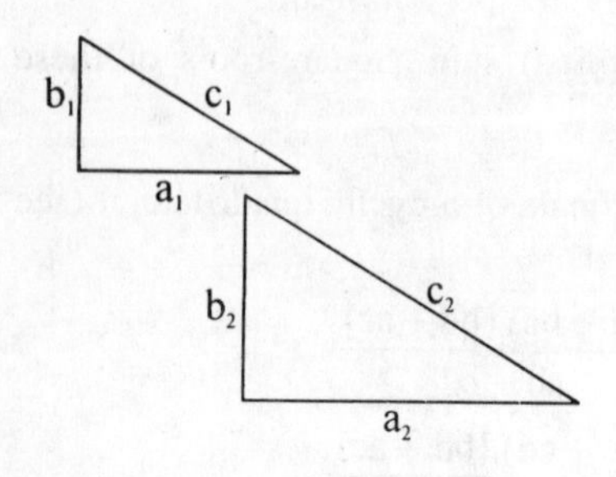

Comment: Take two right triangles of (generally integral) sides $a_1, b_1, c_1, a_2, b_2, c_2, a_i \leqq b_i \leqq c_i$. Form a quadrilateral with sides $c_1a_2, c_2a_1, c_1b_2, c_2b_1$ (Products of hypotenuse with sides of the other triangle). One diagonal $= a_1a_2 + b_1b_2$. The other diagonal $= a_1b_2 + a_2b_1$.

NOTE: Bhāskarācārya claims that his method is much simpler than Brahmagupta's and wonders how it was not discovered earlier.

Example

Take $a_1 = 3, b_1 = 4, c_1 = 5, a_2 = 5, b_2 = 12, c_2 = 13$. Then the quadrilateral has sides 52, 39, 25, 60 which is the same one in stanza (CXCV). With sequence of sides (25, 39, 60, 52), the diagonals are

$5 \times 3 + 12 \times 4 = 63.$

$12 \times 3 + 5 \times 4 = 56.$

However, with sequence of sides (25, 60, 52, 39), the smaller diagonal is the same and the bigger diagonal is the product of the hypotenuse of the two optional right triangles, that is, the bigger diagonal is $c_1 c_2 = 65$.

There are no square roots nor big products.

Bhāskarācārya's formula can be proved by using Brahmagupta's formula and substituting $c_1a_2, c_2a_1, c_1b_2, c_2b_1$ for the sides and nothing $c_1^2 = a_1^2 + b_1^2$ and $c_2^2 = a_2^2 + b_2^2$.

REMARK: Is is evident that if four sides of a cyclic quadrilateral are given then, to compute the diagonals, one needs two optional right triangles. Suppose we consider two right triangles with sides 3, 4, 5 and 8, 15, 17. Then, by the above method, diagonals of the quadrilateral with sequence of sides (25, 39, 60, 52) should be

$3 \times 8 + 4 \times 15 = 84$ and $3 \times 15 + 4 \times 8 = 77$.

These are against the correct values 63, 56. Thus these right triangles are not compatible to a quadrilateral with sides 25, 39, 60, 52 (taken in any sequence), and we cannot choose two right triangles in an arbitrary manner. A pair of compatible right triangles to a cyclic quadrilateral can be choosen in the following manner:

Divide or multiply the smallest and biggest sides of the quadrilateral (hopefully) by any (convenient) positive number. The outcomes are the perpendiculars of one optional right triangle. Now divide the remaining two sides of the quadrilateral by the hypotenuse of the first optional triangle. The quotients are the perpendicular sides of the other optional right triangle.

Example

क्षेत्रे यत्र शतत्रयं क्षितिमितिस्तत्त्वेन्दुतुल्यं मुखं
बाहू खोत्कृतिभिः शरातिधृतिभिस्तुल्यौ च तत्र श्रुती।
एका खाष्टयमैः समा तिथिगुणैरन्याथ तल्लंबकौ
तुल्यौ गोधृतिभिस्तथाजिनयमैर्योगाच्छ्रवोर्लम्बयोः।। CXCIX ।।

तत्खंडे कथयाधरे श्रवणयोर्योगाच्च लम्बाबधे
तत्सूची निजमार्गवृद्धभुजयोर्योगाद्यथा स्यात्तथा।
साबाधं वद लंबकं च भुजयोः सूच्याः प्रमाणे च के
सर्वं गाणितिक प्रचक्ष्व नितरां क्षेत्रेऽत्र दक्षोऽसि चेत्।। CC ।।

A (cyclic) quadrialteral has base 300, upper side 125, right side 260, left side 195 and the diagonals are 280 and 315. The perpendiculars are 189 and 224. Find the lengths of the following:

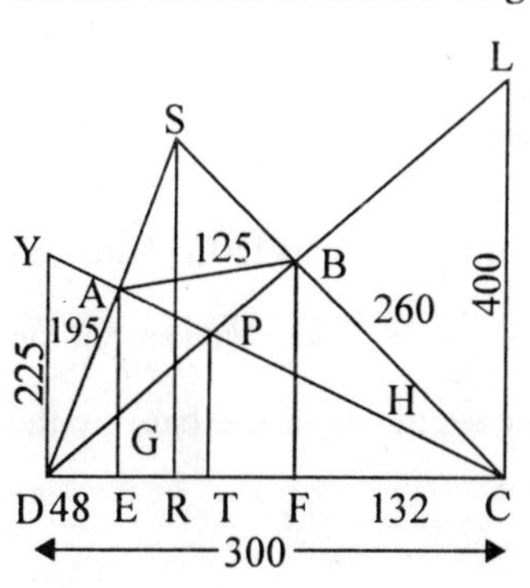

(1) GE, HF, DE, EC, DF, FC
(2) PT, DT, TC
(3) SR, DR, RC
(4) SD, SC.

Comment: DC = 300, CB = 260, AB = 125, AD = 195. Other values, AC = 315, BD = 280, AE = 189, BF = 224, can be computed from the above and need not have been given.

$DE^2 = (195)^2 - (189)^2 \quad \therefore DE = 48, EC = 252$

$CF = \sqrt{260^2 - 224^2} = 132, FD = 168, EF = 120$.

From similar triangles AEC, YDC, $\frac{CE}{AE} = \frac{DC}{DY}$ $\therefore$ DY = 225.

Similarly, we get RC = 300 – DR, 33 SR = 56 (300 – DR),

16 SR = 63 DR. Solving RC = $\frac{3564}{17}$. GE = 64, GD = 80, HF = 99,

HC = 165, RF = $77\frac{11}{17}$. SA = $\frac{195 \times 15}{17}$, SB = $\frac{2600}{17}$.

PT = 144, CT = 192.

लंबतदाश्रितबाह्वोर्मध्यं संध्याख्यमस्य लंबस्य।
संध्यूना भूः पीठं साध्यं यस्याधरं खण्डम्।। CCI ।।

तत्संधिद्विष्ठः परलंबश्रवणाहतोऽन्यपीठेन।
भक्तो लंबश्रुत्योर्योगात् स्यातामधः खण्डे।। CCII ।।

(Here Bhāskarācārya gives his method to solve the above intricate example.) DE and FC (projections of AD and BC on the base) are called *sandhis.* When these are subtracted from the base, the remainders are called *pīṭha* or complements of the projections.
Then

$$GE = \frac{DE \times BF}{DF}, GD = \frac{DE \times BD}{DF}.$$

लंबौ भूघ्नौ निजनिजपीठविभक्तौ च वंशो स्तः।
ताभ्यां प्राग्वत् श्रुत्योर्योगाल्लंबः कुखण्डे च।। CCIII ।।

YD and LC are called *'poles'.* They are at right angles to the base and their tops are on the diagonals. YD = $\frac{DC \times AE}{EC}$. Then we can get the height of the point of intersection and the segments of the base by the use of stanza CLXVII.

लंबहृतो निजसंधिः परलंबगुणः समाह्वयो ज्ञेयः।
समपरसन्ध्योरैक्यं हारस्तेनोद्धृतौ तौ च।। CCIV ।।

समपरसन्धी भूघ्नौ सूच्याबाधे पृथक् स्याताम्।
हारहृतः परलंबः सूचीलंबो भवेद्भूघ्नः।। CCV ।।

सूचीलंबघ्नभुजौ निजनिजलंबोद्धृतौ भुजौ सूच्याः।
एवं क्षेत्रक्षोदः प्राज्ञैस्त्रैराशिकात् क्रियते।। CCVI ।।

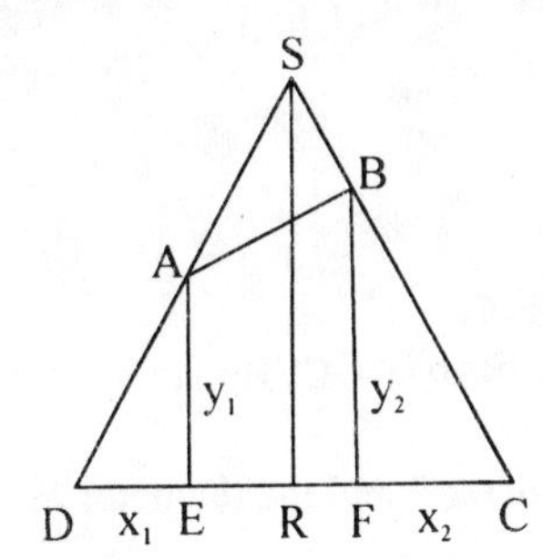

$$DR = \frac{DC \times y_2 \times x_1}{y_2x_1 + y_1x_2} = \frac{DC \times \frac{y_2}{y_1} \times x_1}{y_2 \times \frac{x_1}{y_1} + x_2}.$$

$$SR = \frac{DC \times y_2}{x_2 + \frac{y_2x_1}{y_1}}.$$

Comment: Two equations (for DR and SR) derived in stanza CC are given here by Bhāskarācārya in different forms.

Formula to compute the circumference of a circle given its diameter

व्यासे भनंदाग्निहते विभक्ते खबाणसूर्यैः परिधिस्तु सूक्ष्मः।
द्वाविंशतिघ्ने विहृतेऽथ शैलैः स्थूलेभोऽथवा स्याद्व्यवहारयोग्यः।। CCVII ।।

For a given circle, the product of the diameter and 3927/1250 gives a good approximate value of the circumference, while the product of the diameter and 22/7 gives a rough approximation of the circumference.

Comment: Circumference $\doteq \frac{3927}{1250} \times$ Diameter

$\doteq \frac{22}{7} \times$ Diameter.

It is known that $\frac{\text{circumference}}{\text{diameter}}$ is a transcendental irrational number and is usually denoted by π . π is not a root of any algebraic equation. For calculations we use approximate values $\frac{22}{7}$, $\frac{3927}{1250}$, 3.1416, etc. It shows that Bhāskara and his (Indian) foremathematicians were aware of the irrationality of $\left(\frac{\text{circumference}}{\text{diameter}}\right)$, that is that of π. The Vedic term for π is *trita* (*Ṛgveda* 1.52.5).

Example

विष्कंभमानं किल सप्त यत्र तत्र प्रमाणं परिधेः प्रचक्ष्व।
द्वाविंशतिर्यत्परिधिप्रमाणं तद्व्याससंख्या च सखे विचिन्त्य।। CCVIII ।।

Find the circumference of a circle of diameter 7 and the diameter of a circle whose circumference is 22.

Comment: Circumference = $\frac{22}{7} \times 7 = 22$.

$$\text{Diameter} = 22 \div \frac{22}{7} = 7.$$

This formula requires calculus for its proof.

Formulae: Area of a disk, surface area and volume of a sphere

वृत्तक्षेत्रे परिधिगुणितव्यासपादः फलं यत्
क्षुण्णं वेदैरुपरि परितः कंदुकस्येव जालम्।
गोलस्यैवं तदपि च फलं पृष्ठजं व्यासनिघ्नं
षड्भिर्भक्तं भवति नियतं गोलगर्भे घनाख्यम्।। CCIX ।।

Area of a disk = $\frac{1}{4}$ (circumference) × (diameter). Area of the surface of a sphere = 4 × area of its great circle. Volume of a sphere = $\frac{1}{6}$ (surface area) × (diameter).

Comment: If the radius of the sphere is r,

Area of great circle = $\frac{1}{4}$ $(2\pi r)$ $(2r) = \pi r^2$. Area of surface = $4\pi r^2$.

Volume of the sphere = $\frac{1}{6}$ $(4\pi r^2)$ $(2r) = \frac{4}{3}$ πr^3.

These formulae are proved in any calculus text.

Example

यद्व्यासस्तुरगैर्मितः किल फलं क्षेत्रे समे तत्र किं
व्यासः सप्तमितश्च यस्य सुमते गोलस्य तस्याऽपि किम्।
पृष्ठे कंदुकजालसन्निभफलं तस्यैव गोलस्य किं
मध्ये ब्रूहि धनं फलं च विमलां चेद्वेत्सि लीलावतीम्।। CCX ।।

What is the area of a disk of diameter 7? What is the area of a net which just encloses a ball of diameter 7? What is its volume? O wise friend, answer if you know pellucid Līlāvatī well.

Comment: Area of the disk = $\frac{22}{7} \times \left(\frac{7}{2}\right)^2 = \frac{77}{2} = 38.5$ sq. units.

$$\text{Surface area} = 4 \times \frac{22}{7} \times \frac{49}{4} = 154 \text{ sq. units.}$$

$$\text{Volume} = \frac{4}{3} \times \frac{22}{7} \times \frac{7^3}{2^3} = \frac{539}{3} = 179\frac{2}{3} \text{ cu. units.}$$

व्यासस्य वर्गे भनवाग्निनिघ्नं सूक्ष्मं फलं पंचसहस्त्रभक्ते।
रुद्राहते शक्रहृतेऽथवा स्यात् स्थूलं फलं संव्यवहारयोग्यम्।। CCXI ।।

For greater accuracy,

Area of a disk = $\frac{3927}{5000}$ (Diameter)^2, and

for most practical purposes this may be $\frac{11}{14}$ (Diameter)^2.

घनीकृतव्यासदलं निजैकविंशांशयुग्गोलफलं घनं स्यात्।। CCXII ।।

Comment: Volume $V = \frac{D^3}{2} + \frac{D^3}{2} \times \frac{1}{21}$ where D = Diameter.

$$V = \frac{22}{2 \times 21} D^3 \doteq \frac{1}{6} \pi 8r^3 = \frac{4}{3} \pi r^3.$$

Formula: Perpendicular to a chord

ज्याव्यासयोगान्तरघातमूलं व्यासस्तदूनो दलितः शरः स्यात्।
व्यासाच्छरोनात् शरसंगुणाच्च मूलं द्विनिघ्नं भवतीह जीवा।। CCXIII ।।

जीवार्धवर्गे शरभक्तयुक्ते व्यासप्रमाणं प्रवदन्ति वृत्ते।। CCXIV ।।

[These two stanzae give three relations between chord, its arrow and diameter of a circle. Indeed, one of the three is computed when the other two are known. BC = chord, MB = radius r, AE = Diameter 2r, AD = 'Arrow'.] Subtract the (postive) square-root of the product of the sum and difference of the (given) diameter and the (given) chord. The result divided by two is the arrow.

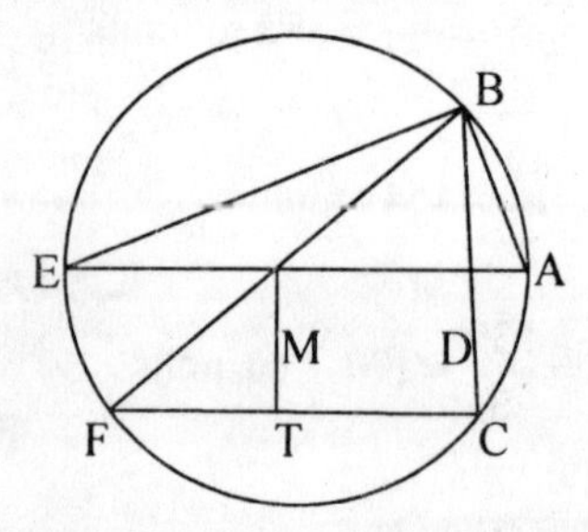

Twice the (positive) square-root of the product of the (given) arrow and the difference of the (given) diameter and the arrow is the chord.

Divide the square of half the (given) chord by the given arrow. The sum of this result and the arrow is the diameter.

[Indeed, $AD = \frac{1}{2}[AE - \sqrt{(AE+BC)(AE-BC)}$

$$BC = 2BD = 2\sqrt{AD(AE-AD)}$$

$$AE = AD + \frac{(BD)^2}{AD}.]$$

Comment: $FC^2 = BF^2 - BC^2 = AE^2 - BC^2 = (AE+BC)(AE-BC)$

$$TC = \frac{1}{2}FC \quad \therefore AD = AM - MD$$

$$= \frac{1}{2}AE - \frac{1}{2}\sqrt{(AE+BC)(AE-BC)}$$

$$BD^2 = AD \times DE = AD(AE-AD)$$

$$\therefore BC = 2BD = 2\sqrt{AD(AE-AD)}$$

$$ED \times AD = BD^2$$

$$\therefore ED = \frac{BD^2}{AD} \quad \therefore AE = AD + \frac{BD^2}{AD}.$$

Example

दशविस्तृतिवृत्तान्तर्यत्र ज्या षण्मिता सखे।
तत्रेषुं वद बाणाज्ज्यां ज्याबाणाभ्यां च विस्तृतिम्।। CCXV ।।

O friend, diameter of a circle is 10 and its chord is 6. Find the length of the arrow. If the arrow is given, tell (me) the length of the chord. If the arrow and the chord are given, find the diameter.

Comment: AE = 10, BD = DC = 3.

$$MD = \sqrt{MB^2 - BD^2} = \sqrt{5^2 - 3^2} = 4 \quad \therefore AD = 1.$$

In the second example, $BD^2 = BM^2 - MD^2 = 9 \quad \therefore BD = 3$ and BC = 6.

In the third example, $AE = AD + \frac{(BD)^2}{AD} = 1 + \frac{9}{1} = 10.$

Formula: To find the sides of a regular inscribed polygon

त्रिद्व्यंकाग्निनभश्चन्द्रैस्त्रिबाणाष्टयुगाष्टभिः।
वेदाग्निबाणखाश्वैश्च खखाभ्राभ्ररसैः क्रमात्।। CCXVI ।।

बाणेषुनखबाणैश्च द्विद्विनंदेषुसागरैः।
कुरामदशवेदैश्च वृत्ते व्याससमाहते।। CCXVII ।।

खखखाभ्रार्कसंभक्ते लभ्यन्ते क्रमशो भुजाः।
वृत्तान्तत्र्यस्त्रपूर्वाणाम् नवास्त्रान्तं पृथक् पृथक्।। CCXVIII ।।

Sides of regular insribed polygons are:

No. of sides	Side
3	103923a
4	84853a
5	70534a
6	60000a
7	52055a
8	45922a
9	41031a

where $a = \frac{\text{Diameter}}{120000}$.

Comment: For a regular polygon of n sides inscribed in a circle of radius r, side = $2 r \sin \frac{\pi}{n}$.

Using Trigonometrical tables, we get different values for n = 3, 4, 5, ..., 9.

अथ स्थूलजीवाज्ञानार्थं लघुक्रिया करणसूत्रं वृत्तम्

Formula for the length of a chord: a rough approximation

चापोननिघ्नपरिधिः प्रथमाह्वयः स्यात्
पंचाहतः परिधिवर्गचतुर्थभागः।
आद्योनितेन खलु तेन भजेच्चतुर्घ्नं।
व्यासाहतं प्रथमप्राप्तमिह ज्यका स्यात्।। **CCXIX** ।।

Subtract the (given) arc (cut off by the chord) from the circumference. Multiply the remainder by the circumference. Call (this) product 'First'. (Now) multiply the square of the circumference by 5/4, subtract from it the First, and divide the First by this remainder. The result multiplied by four times the diameter is the (length of the) chord.

Comment: d = diameter, p = circumference, c = arc length.

$$\text{Length of the chord} = \frac{4dc\,(p-c)}{\frac{5}{4}p^2 - c\,(p-c)}$$

This formula gives an approximate value and it is difficult to trace its derivation. When the arc length c is zero, so is the chord length. The same is true if c = p, the circumference. So length of the chord should be proportional to c/(p – c). The third factor should be a multiple of the diameter d. Clearly the denominator must be of the second degree (one less than that of the numerator) in p and c. So the formula must be $\ell = \frac{xc\,(p-c)}{y - c\,(p-c)}$. Now we have two special cases:

(i) $\ell = p$ when $c = \frac{p}{2}$ and

(ii) $\ell \doteq \frac{p}{2}$ when $c = \frac{p}{6}$.

Solving these two equations for x and y we get the result.

Example

अष्टादशांशेन समानवृत्तम् एकादिनिघ्नेन च यत्र चापम्।
पृथक् पृथक् तत्र वदाशु जीवाम् खार्कैर्मितं व्यासदलं च यत्र।। CCXX ।।

Find the length of the chord of a circle whose radius is 120 and arc one-eighteenth of its circumference. Find the lengths when the arc length is doubled, trebled etc.

Comment: Here d = 240, $c = \frac{1}{18}p$.

$$\ell \doteq \frac{4d\frac{1}{18}p \times \frac{17p}{18}}{\frac{5p^2}{4} - \frac{17p^2}{324}} = \frac{68d}{388} = \frac{17 \times 240}{97} = 42.06.$$

Its exact value = d *sin* 10° = 240 × (0.1736) = 41.664.

When $c = \frac{2}{18}p$, $\ell = 84.12$.

In Bhāskarācārya's time (A.D. 1150), there were no tables of logarithms or trigonometric functions and so the students were given approximate formulae.

Formula to find the arc length, given the chord

व्यासाब्धिघातयुतमौर्विकयाविभक्तो
जीवांघ्रिपंचगुणितः परिधेस्तु वर्गः।
लब्धोनितात् परिधिवर्गचतुर्थभागात्
आप्ते पदे वृतिदलात्पतिते धनुः स्यात्।। CCXXI ।।

Divide the product of the square of the circumference and 5/4 times the (given) chord by the sum of the chord and four times the diameter. Subtract this quotient from one-fourth of the square of the circumference. The (positive) square-root of this remainder subtracted from half the circumference gives the bow (the smaller arc).

Comment: Arc length $c = \frac{p}{2} \pm \sqrt{\frac{p^2}{4} - \frac{\frac{5}{4}p^2\ell}{4d+\ell}}$.

Solve the above formula (cf. stanza CCXX) for c and the two answers give the major arc and the minor arc respectively. (Bhāskarācārya intends to give only the minor arc.)

Example

विदिता इह ये गुणास्ततो वद तेषामधुना धनुर्मितिम् ।
यदि तेऽस्ति धनुर्गुणक्रियागणिते गाणितिकातिनैपुणम् ।। CCXXII ।।

O arithmetician! If you know the formula for the arc length of a circle, find what proportion the arc length is of the circumference when the lengths of the chords are 42, 82, 120, ... The diameter of the circle is 240.

Comment: Here $\ell = 42$, $d = 240$, $p = 754.16$. Since p is large, to facilitate the calculations, we compute

$$\frac{c}{p} = \frac{1}{2} \pm \sqrt{\frac{1}{4} - \frac{\frac{5}{4}\ell}{4d + \ell}} \doteq \frac{17}{18} \text{ or } \frac{1}{18} \text{ for } \ell = 42.$$

If $\ell = 82$, $\frac{c}{p} = \frac{1}{9}$ or $\frac{8}{9}$.

CHAPTER 29

Volume

गणयित्वा विस्तारं बहुषु स्थानेषु तद्युतिर्भाज्या।
स्थानकमित्या सममितिरेवं दैर्ध्येच वेधे च
क्षेत्रफलं वेधगुणं खाते घनहस्तसंख्या स्यात्॥ **CCXXIII** ॥

If it is an irregular ditch (or solid), measure the breadth at various points, add them and divide by the number of points. That is the (average) breadth. Similarly, calculate the (average) length and the (average) depth (or height). The product of the three (averages) will give the (average) volume.

Example

भुजवक्रतया दैर्घ्यं दशेशार्ककरैर्मितम्।
त्रिषु स्थानेषु षट्पंचसप्तहस्ता च विस्तृतिः॥ **CCXXIV** ॥
यस्य खातस्य वेधोऽपि द्विचतुस्त्रिकरः सखे।
तत्र खाते कियन्तः स्युर्घनहस्ताः प्रचक्ष्व मे॥ **CCXXV** ॥

O friend! Measurements of an irregular solid were:

lengths 10, 11, 12 C (cubits), breadths 6, 5, 7 C and heights 2, 4, 3 C. Find its volume.

Comment: Average length $= \frac{10+11+12}{3} = 11\,C$.

Average breadth $= \frac{6+5+7}{3} = 6\,C$.

Average height $\frac{2+3+4}{3} = 3\,C$.

Average volume $= 11 \times 6 \times 3 = 198\,C^3$.

It should be noted that the measurements should be carried out at equal intervals. It is clear that Bhāskarācārya gives practical methods.

Formula: Volume of a pyramid and its frustum

मुखजतलजतद्युतिजक्षेत्रफलैक्यं हृतं षड्भिः।
क्षेत्रफलं सममेतद्वेधगुणं धनफलं स्पष्टम्।
समखातफलत्र्यंशः सूचीखाते फलं भवति।। CCXXVI।।

Find the sum of the areas of the mouth (top), bottom (base) and the rectangle whose (adjacent) sides are sums of the lengths and breadths (of the mouth and the bottom). The sum (of the three areas) divided by six is mean area of ditch (frustum). The mean area multiplied by depth (height) is the volume. (This) volume divided by three is volume of pyramid.

Comment: Volume of frustum

$$= \frac{1}{6}\ MM'\ [pq + p'q' + (p + p')\ (q + q')].$$

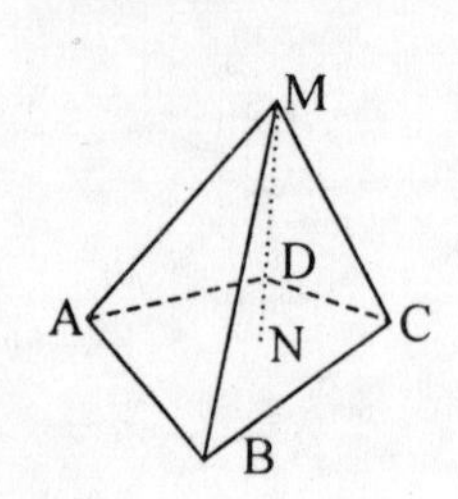

Volume of a pyramid is $\frac{1}{3}$ that of a prism with the same base and same height.

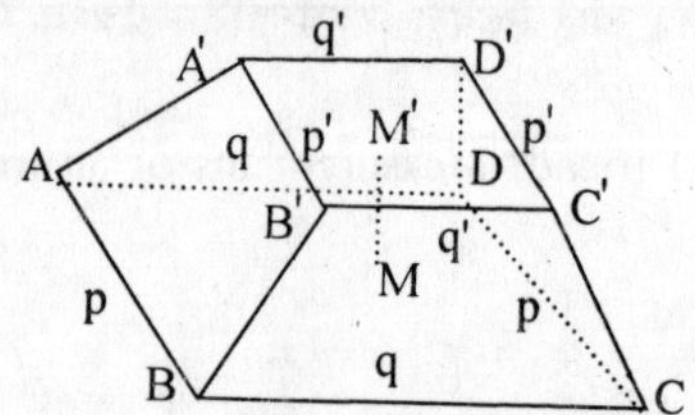

These are standard formulae given in texts.
Here ABCD and A'B'C'D' are similar rectangles.

Example 1

मुखे दशद्वादशहस्ततुल्यं विस्तारदैर्घ्यं तु तले तदर्धम्।
यस्याः सखे सप्तकरश्च वेधः का खातसंख्या वद तत्र वाप्याम्।। CCXXVII।।

There is a well in the shape of a frustum of a pyramid. (The base is smaller than the top.) Its top is a rectangle of sides 10×12 C^2 (cubits) and the base is half the size of the top, i.e. 5×6 C^2. If its height is 7 C, O friend, find the volume of the well.

Comment: Volume $= \frac{7}{6}$ $[120 + 30 + 15 \times 18]$
$= 490$ C^3.

Examples 2, 3, 4, 5

खातेऽथ तिग्मकरतुल्यचतुर्भुजे च किं स्यात्फलं नवमितः खलु यत्र वेधः।
वृत्ते तथैव दशविस्तृतिपंचवेधे सूचीफलं वद तयोश्च पृथक् पृथङ्मे।।CCXXVIII।।

(2) The top of a hole is 12×12 C^2 (cubits) and its depth is 9 C. Find its volume.
(3) The circular top of a hole is of diameter 10 C and its depth is 5 C. Find its volume. Find the volume of a pyramid (cone) whose base and height are: (4) as in (2); (5) as in (3).

Comment: (2) This is a box (rectangular parallelo-piped) whose volume $= 12 \times 12 \times 9 = 1296$ C^3.

(3) This is a right circular cylinder whose volume $= \frac{22}{7} \times (5)^2 \times 5$

$= \frac{2750}{7} = 392.86$ C^3.

(4) Volume of the pyramid $= \frac{1}{3} . 1296 = 432$ C^3.

(5) Volume of the cone $= \frac{1}{3}$ $(392.86) = 130.95$ C^3.

Formula: Volume of a prism

उच्छ्रयेन गुणितं चितेः किल क्षेत्रसंभवफलं घनं भवेत्।
इष्टिकाघनहृते घने चितेः इष्टिकापरिमितिश्च लभ्यते।। CCXXIX ।।
इष्टिकोच्छ्रयहृदुच्छ्रितिश्चितेः स्युःस्तराश्च दृषदां चितेरपि।। CCXXX।।

Volume of a prism equals the product of the area of its base and height. Number of (rectangular parallelo-piped) bricks in a heap (prism) is obtained by dividing the volume of the heap by the volume of one brick. If the height of the heap is divided by the height of a brick, we get the number of horizontal layers of bricks.

Comment: This is straight forward.

Example

अष्टादशांगुलं दैर्घ्यं विस्तारो द्वादशांगुलः।
उच्छ्रितिस्त्र्यंगुला यासामिष्टिकास्ताश्चितौ किल।। CCXXXI ।।
यद्विस्तृतिः पंचकराष्टहस्तं दैर्घ्यं च यस्यां त्रिकरोच्छ्रितिश्च।
तस्यां चितौ किं फलमिष्टिकानां संख्या च का ब्रूहि कति स्तराश्च।। CCXXXII।।

A rectangular prism $8 \times 5 \times 3\ C^3$ (C = cubits) is to be constructed by using bricks of dimensions $18 \times 12 \times 3\ A^3$ (A = *aṅgulas*). Find the volume of the prism, number of bricks required and the number of layers. (1 C = 24A).

Comment: Volume of the prism $= 8 \times 5 \times 3 = 120\ C^3$

$$\text{Volume of a brick} = \frac{18 \times 12 \times 3}{24 \times 24 \times 24} = \frac{3}{64}\ C^3$$

$$\text{No. of bricks} = 120 \div \frac{3}{64} = 2560$$

$$\text{No. of layers} = 3 \div \frac{3}{24} = 24.$$

CHAPTER 30

Wood Cutting

पिंडयोगदलमग्रमूलयोः दैर्ध्यसंगुणितमंगुलात्मकम् ।
दारुदारणपथैः समाहतं षट्स्वरेषुविहृतं करात्मकम् ।। CCXXXIII ।।

[Log is a frustum of a cone. AB = d_1, DC = d_2, MN = h.]

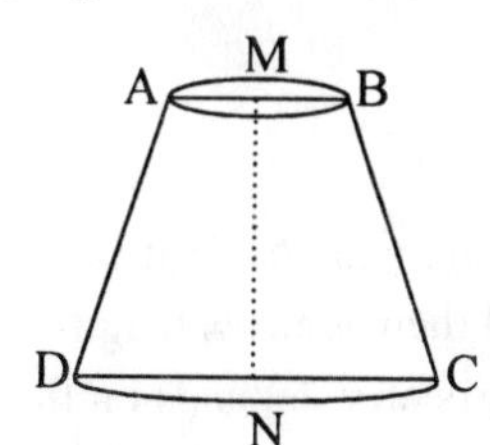

Area of a section (ABCD) = $\frac{h}{2}$ (d_1 + d_2).

Multiply this by the number of times the wood is cut to get the total area in A^2, (A = *aṅgulas*). To get the answer in C (cubits) divide by 576.

Comment: Section ABCD is a trapezium. If several sections are made, obviously they are not equal in area. So this gives an approximate answer. 1 $C^2 = 24 \times 24\ A^2 = 576\ A^2$.

Example

मूले नखांगुलमितोऽथ नृपांगुलोऽग्रे
पिण्ड: शतांगुलमितं किल यस्य दैर्ध्यम्।
तद्दारुदारणपथेषु चतुर्षु किं स्यात्
हस्तात्मकं वद सखे गणितं द्रुतं मे।। CCXXXIV ।।

A log of wood has base 20 A (*aṅgulas*) long, top 16 A long and height 100 A. If it is cut at four places, O friend, find the total surface area of the wooden pieces.

Comment: Area $= 4 \times \frac{100}{2} (20 + 16) \frac{1}{576} C^2$
$= 12.5\ C^2$.

Formula: oblique cutting of a log

छिद्यते तु यदि तिर्यगुक्तवत् पिंडविस्तृतिहते: फलं तदा।
इष्टिकाचिति दृषच्चितिखातक्राकचव्यवहृतौ खलु मूल्यम्।
कर्मकारजनसंप्रतिपत्त्या तन्मृदुत्वकठिनत्ववशेन।। CCXXXV ।।

A log of wood is such that its vertical sections are squares. If it is cut obliquely, then the sections are rectangular and their areas = length × breadth. Cost of a heap of bricks or stones or carpentry depends on the number and skill of workers as well as softness or hardness of wood.

Example

यद्विस्तृतिर्दन्तमितांगुलानि पिंडस्तथा षोडश यत्र काष्ठे।
छेदेषु तिर्यङ्नवसु प्रचक्ष्व किं स्यात्फलं तत्र करात्मकं मे।। CCXXXVI।।

A log is 32 A (*aṅgulas*) long and 16 A thick. It was cut obliquely at 9 places. Find the total area of the sections.

Comment: Area $= \frac{9 \times 16 \times 32}{576} = 8C^2$

CHAPTER 31

Volume of a Heap of Grain

अनणुषु दशमांशोऽणुष्वथैकादशांशः
परिधिनवमभागः शूकधान्येषु वेधः।
भवति परिधिषष्ठे वर्गिते वेधनिघ्ने
घनगणितकराः स्युर्मागधा ताश्च खार्य्यः।। CCXXXVII।।

If the grains are big and spherical, their heap (right circular cone) has a height which is $\frac{1}{10}$ th of the circumference. If they are small and spherical, the proportion is $\frac{1}{11}$ th and if they are pointed, it is $\frac{1}{9}$ th. Volume can be computed by (height) $\times \left(\frac{\text{circumference}}{6}\right)^2$. This gives the answer in C^3 (cubits); it also represents the number of Magadha *khārikās*.

Comment: A heap of grains is in the form of a right circular cone. Its

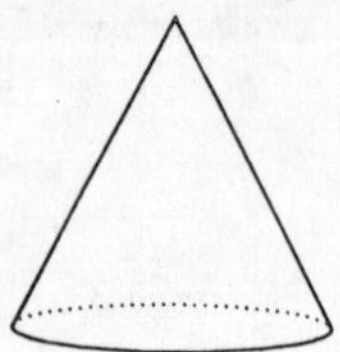

circumference can be measured but not its height without disturbing the heap. So Bhāskarācārya gives methods to compute the heights for 3 different types of grains.

Volume = $\frac{1}{3}\pi r^2 h$ (r radius) as seen before. Bhāskarācārya's formula

$= \left(\frac{2\pi r}{6}\right)^2 h = \frac{\pi^2 r^2 h}{9} \doteq \frac{1}{3}\pi r^2 h$ if we take $\pi = 3$. This is good for all practical purposes for the village folk.

Example

समभुवि किल राशिर्यः स्थितः स्थूलधान्यः
परिधिपरिमितिर्भो हस्तषष्टिर्यदीया।
प्रवद गणक खार्यः किंमिताः सन्ति तस्मिन्
अथ पृथगणुधान्ये शूकधान्ये च शीघ्रम्।। CCXXXVIII ।।

On a flat ground there is a heap of big spherical grains of perimeter 60 C (cubits). O mathematician, find how many *khārikās* (K) are there. Find the same if the grains are small or pointed.

Comment: (i) Big, spherical: Circumference p = 60 C, h = 6 C, Volume $= \left(\frac{60}{6}\right)^2 6 = 600$ K;

(ii) small spherical: $h = \frac{60}{11}$, Volume $100 \times \frac{60}{11} = 545\frac{5}{11}$ K;

(iii) pointed: $h = \frac{60}{9}$, Volume $= 100 \times \frac{60}{9} = 666\frac{2}{3}$ K.

Formula: Heap in a corner

द्विवेदसत्रिभागैकनिघ्नात्तु परिधेः फलम्।
भित्त्यन्तर्बाह्यकोणस्थराशेः स्वगुणभाजितम्।। CCXXXIX ।।

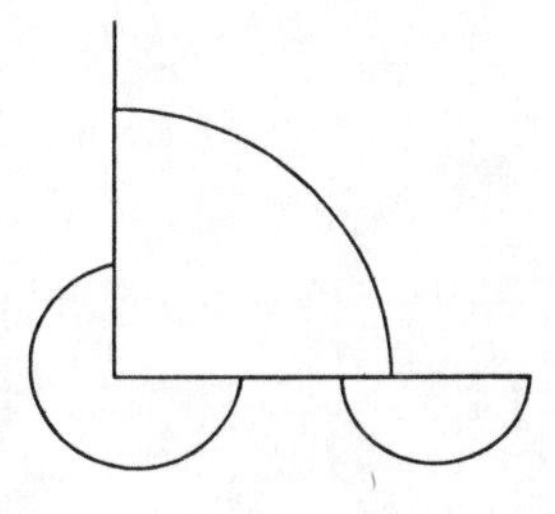

[If the heap is in an outer corner, the base is $\frac{3}{4}$ th area of the disk. If it is against one wall, the area is half. If it is between the two walls, the area is one-fourth.] If the heap is kept in the outer corner, the circumference of the circle (of which the base of the heap is a part) is $\frac{4}{3}$ of the length of the base. If the heap is on one side of a wall, multiply by 2 and if it is in the interior it will be 4 times. The actual volumes of the heaps of grain will be $\frac{3}{4}, \frac{1}{2}, \frac{1}{4}$ respectively of the volumes so obtained.

Example

परिधिर्भित्तिलग्नस्य राशेस्त्रिंशत्करः सखे
अन्तःकोणस्थितस्यापि तिथितुल्यकरः किल।
बहिःकोणस्थितस्यापि पंचघ्ननवसंमितः
तेषामाचक्ष्व मे क्षिप्रं घनहस्तान् पृथक् पृथक्।। **CCXL** ।।

Find the volumes of heaps of grain stored in
(i) front corner with length of the base ℓ = 45 C (cubits),
(ii) on a wall ℓ = 30 C,
(iii) inside corner ℓ = 15 C.

Comment: In all cases circumference = 60 C. Let h be the height (determined according to stanza CCXXXVII).

(i) $\text{Volume} = \frac{3}{4}\left(\frac{60}{6}\right)^2 \text{h} = \frac{3}{4}\ 100\ \text{h} = 75\ \text{h}\ \text{C}^3$

(ii) $\text{Volume} = \frac{1}{2} \times 100 \times \text{h} = 50\ \text{h}\ \text{C}^3$

(iii) $\text{Volume} = \frac{1}{4} \times 100 \times \text{h} = 25\ \text{h}\ \text{C}^3$

CHAPTER 32

Shadows

Formula: Length of a shadow

छाययोः कर्णयोरन्तरे ये तयोः वर्गविश्लेषभक्ता रसाद्रीषवः।
सैकलब्धेः पदघ्नं तु कर्णांतरम् भांतरेणोनयुक्तं दले स्तः प्रभे।। CCXLI।।

Obtain squares of the (given) difference of hypotenuses and that of shadows. Divide 576 by the difference of these two squares. Add one to this quotient and then take the (positive) square-root. Multiply this result by the difference of hypotenuses, and write the product in two places. Add the difference of shadows to one place, and subtract the same from the other place. The outcomes divided by two are lengths of shadows.

[Height of the vertical pole = r, its shadows are t, t + q and the corresponding hypotenuses are c, c + p. We are given r, p, q and we have to find t when q > p.

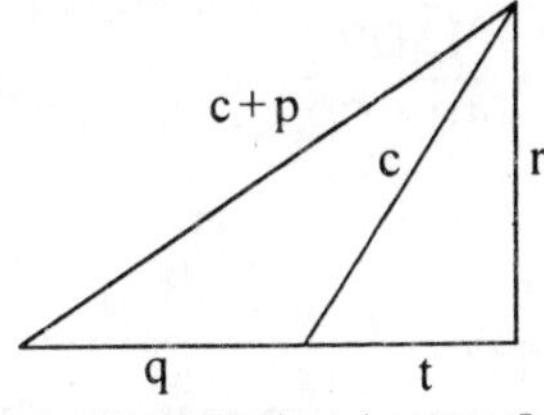

$$t = \frac{1}{2}\left[-q + p\sqrt{1 + \frac{576}{q^2 - p^2}}\right]$$

The other shadow is t + q.]

Comment: This concerns the altitude of the Sun. Some calculations are required. A pole (of length r) is fixed and a string is tied to its top. Marks are made on the ground where the shadows fall and the corresponding lengths of the string are noted.

$$(c+p)^2 = r^2 + (q+t)^2 \text{ and } c^2 = r^2 + t^2$$

$$\therefore 2cp = 2qt + m \text{ where } m = q^2 - p^2$$

$$\therefore 4c^2p^2 = (2qt+m)^2$$

$$\therefore 4p^2(r^2+t^2) = 4q^2t^2 + 4qtm + m^2$$

$$\therefore 4t^2(q^2-p^2) + 4mqt + m^2 - 4p^2r^2 = 0.$$

Solve this quadratic equation and choose the root with positive sign and put r = 12. In *Līlāvatī,* height r usually corresponds to a gnomon of 12 *aṅgulas.*

Formula: To find the length of the shadow when the heights of the lamp and pole as well as the distance between them are given.

शंकुः प्रदीपतलशंकुतलांतरघ्नः ।
छाया भवेत् विनरदीपशिखौच्च्यभक्तः ।। CCXLII ।।

[Given AB, DC = 12, BD, find PD.]

$$PD = \frac{12\ DB}{AB - CD}.$$

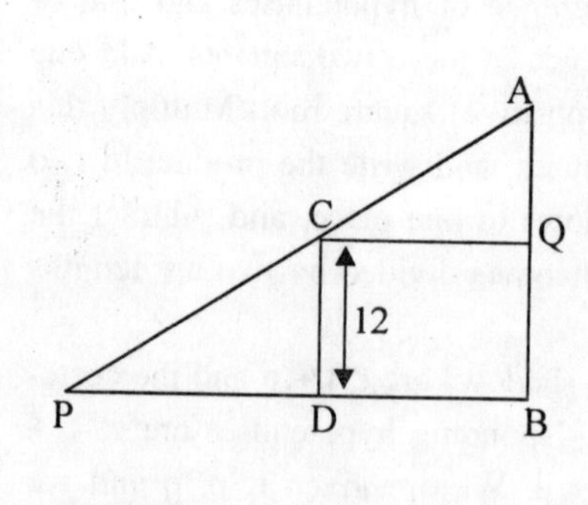

Comment: By similar triangles, PCD, CAQ,

$$\frac{PD}{CD} = \frac{CQ}{AQ} \therefore \frac{PD}{12} = \frac{DB}{AB - BQ}$$

$$\therefore PD = \frac{12\ DB}{AB - CD}.$$

Example

शंकुप्रदीपान्तरभूस्त्रिहस्ता दीपोच्छ्रितिः सार्धकरत्रया चेत् ।
शंकोस्तदाऽर्कांगुलसंमितस्य तस्य प्रभा स्यात्कियती वदाऽऽशु ।। CCXLIII ।।

If BD = 3 C [Cubits = 72 A (*aṅgulas*)], AB = $3\frac{1}{2}$ C (84A) and CD = 12 A. Find the length of the shadow.

Comment: $PD = \frac{12 \times 72}{84 - 12} = 12\ A$.

Formula: to find the height of a lamp

छायोद्धृते तु नरदीपतलांतरघ्ने।
शंकौ भवेन्नरयुते खलु दीपकौच्च्यम्।। CCXLIV ।।

$$AB = \frac{BD \times CD}{PD} + CD.$$

Comment: From the similar triangles PCD, CAQ,

$$\frac{CQ}{AQ} = \frac{PD}{CD} \therefore AQ = \frac{CQ \times CD}{PD} = \frac{BD \times CD}{PD}$$

$$\therefore AB = BQ + AQ = CD + \frac{BD \times CD}{PD}.$$

Example

प्रदीपशंक्वंतरभूस्त्रिहस्ता छायांगुलैः षोडशभिः समा चेत्।
दापोच्छ्रितिः स्यात्कियती वदाऽऽशु प्रदीपशंक्वंतरमुच्यतां मे।। CCXLV ।।

Given BD = 3 C (72A), PD = 16 A, find AB.

Comment: Take CD = 12 A.

$$AB = \frac{72 \times 12}{16} + 12 = 66\ A = \frac{11}{4}\ C.$$

Formula: Distance between the base points

विशंकुदीपोच्छ्रयसंगुणा भा। शंकूद्धृता दीपनरांतरं स्यात्।। CCXLVI ।।

$$BD = \frac{PD\ (AB - CD)}{CD}.$$

छायाग्रयोरंतरसंगुणा भा छायाप्रमाणांतरहृद्‌भवेत् भूः।
भूशंकुघातः प्रभया विभक्तः प्रजायते दीपशिखौच्च्यमेव।। CCXLVII ।।

त्रैराशिकेनैव यदेतदुक्तं व्याप्तं स्वभेदैर्हरिणेव विश्वम्।। CCXLVIII ।।

[AB is the lamp. Pole CD = 12 A is kept at D and D' with lengths of shadows QD and PD'. DD' is also given.] PB – QB = PQ = d

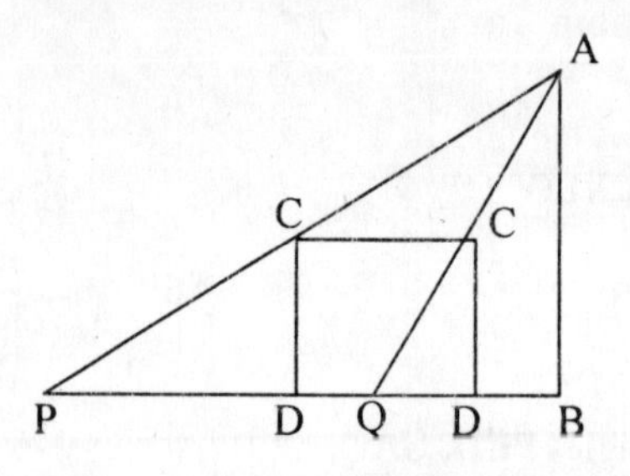

$$PD' - QD = \ell$$

$$QB = \frac{d}{\ell} \times QD$$

$$PB = \frac{d}{\ell} \times PD'$$

$$AB = \frac{QB \times CD}{QD}.$$

All three are obtained by the Rule of Three. Just as the Lord occupies the universe in different forms, the Rule of Three pervades calculations involving shadows.

NOTE: It is not clear as to what purpose is served by calculating the lengths of shadows and the distances of their tips from that of the lamp. Obviously, Bhāskarācārya knew the principle "Light travels in a straight line", but no reference is made to its use in Astronomy.

Example

शंकोर्भार्कमितांगुलस्य सुमते दृष्टा किलाष्टांगुला
छायाग्राभिमुखे करद्वयमिते न्यस्तस्य देशे पुनः।
तस्यैवार्कमितांगुला यदि तदा छाया-प्रदीपान्तरम्
दीपौच्च्यं च कियत् वद व्यवहृतिं छायाभिधां वेत्सि चेत्।। CCXLIX ।।

CD = 12 A, DQ = 8A, DD' = 2C (= 48 A),
PD' = 12A. Find PB, QB and AB.

Comment: QD = 8, DD' = 48, QD' = 40, PD' = 12,

$$PQ = 52.\ BQ = \frac{52 \times 8}{4} = 104\ A$$

$$PB = \frac{52 \times 12}{4} = 156\ A$$

$$AB = \frac{104 \times 12}{8} = 156 \text{ A} .$$

यत्किंचिद्गुणभागहारविधिना बीजेऽत्र वा गण्यते
तत् त्रैराशिकमेव निर्मलधियामेवावगम्यं विदाम्।
एतद्यत् बहुधाऽस्मदादिजडधीधीवृद्धिबुद्ध्या बुधैः
तद्भेदान् सुगमान्विधाय रचितं प्राज्ञैः प्रकीर्णादिकम्।। CCL ।।

Knowledgeable persons are fully aware that examples in Algebra and Arithmetic involving multiplications and divisions are tackled easily by the Rule of Three. To enable mediocre persons like us to master the Rule of Three, they have given us many different examples.

Comment: In this stanza, the importance of the Rule of Three is described. Nowadays 'Unitary System' has replaced the Rule of Three with a consequent loss of speed.
NOTE: It appears that this stanza is not Bhāskarācārya's; it does not seem plausible that he considered himself as mediocre.

CHAPTER 33

Pulverization (Kuṭṭaka)

Before studying Bhāskarācārya's method, let us first see what is 'pulverization'. Take an equation $\frac{ax + b}{c} = y$ where a, b, c are integers. If x, y are also integers, then to find y we have to choose those values of x that will make ax + by a multiple of c. Such an equation is called 'pulverizer'. Pulverize means "beat the problem into powder". Indeed, this indeterminate analysis was designated in Indian Mathematics by the term *kuṭṭaka. Kuṭṭa* means 'to pulverize'.

The above equation in the form ax + b = cy is called Diophantine equation of the first degree. (In this chapter a, b, c, x and y of the equation (ax + b)/c = y stand for dividend, constant, divisor, multiplier and quotient respectively.)

शतं हतं येन युतं नवत्या विवर्जितं वा विहृतं त्रिषष्ट्या।
निरग्रकं स्याद् वद मे गुणं तं स्पष्टं पटीयान् यदि कुट्टकेऽसि।। CCLI।।

O friend, one hundred is multiplied by an integer; 90 is added to or subtracted from the product; the results are exactly divisible by 63. If you are proficient in pulverization, tell me the multiplier correctly. [That is if $100x \pm 90$ is divisible by 63, find x.]

Comment: Let us solve $100x + 90 = 63y$ or

$$100x - 63y = -90 \quad \ldots \quad (1).$$

Let two consecutive solutions be (x_r, y_r), (x_{r+1}, y_{r+1}).
Then $100x_{r+1} - 63y_{r+1} = -90 = 100x_r - 63y_r$
$\therefore 100(x_{r+1} - x_r) = 63(y_{r+1} - y_r)$.
Since $(100, 63) = 1$, $x_{r+1} - x_r = 63$ and $y_{r+1} - y_r = 100$, i.e. (x_r) is an A.P. with C.D. 63 and (y_r) is an A.P. with C.D. 100. Now (1) is $\frac{10x}{9} = \frac{7y}{10} - 1$. Clearly y and $\frac{7y}{10} - 1$ must both be divisible by 10. y = 30 satisfies this. So $x = \frac{9}{10}(20) = 18$. So the solutions are (18, 30), (81, 130), (144, 230), ... Similarly, solutions of $100x - 63y = 90$ are (45, 70), (108, 170), (171, 270), ...

Now let us consider Bhāskarācārya's method and compare it with current ones.

भाज्यो हारः क्षेपकश्चापवर्त्यः केनाप्यादौ संभवे कुट्टकार्थम्।
येन च्छिन्नो भाज्यहारौ न तेन क्षेपश्चैतद् दुष्टमुद्दिष्टमेव।। CCLII।।

परस्परं भाजितयोर्ययोर्यः शेषस्तयोः स्यादपवर्तनं सः।
तेनापवर्तेन विभाजितौ यौ तौ भाज्यहारौ दृढसंज्ञकौ स्तः।। CCLIII।।

मिथो भजेत्तौ दृढभाज्यहारौ यावद् विभाज्ये भवतीह रूपम्।
फलान्यधोधस्तदधो निवेश्य क्षेपस्ततः शून्यमुपान्तिमेन।। CCLIV।।

स्वोर्ध्वे हतेऽन्त्येन युते तदन्त्यं त्यजेन्मुहुः स्यादिति राशियुग्मम्।
ऊर्ध्वो विभाज्येन दृढेन तष्टः फलं गुणो स्यादधरो हरेण।। CCLV।।

एवं तदैवात्र यदा समास्ताः स्युर्लब्धयश्चेद्विषमास्तदानीम्।
यदागतौ लब्धिगुणौ विशोध्यौ स्वतक्षणाच्छेषमितौ तु तौ स्तः।। CCLVI।।

If in a given equation the dividend, divisor and the constant all have a common factor, the equation should be taken as a *proper* one. Dividing by the H.C.F. we get an equation in which the coefficients will have no common factor. If the H.C.F. of the dividend and the divisor does not divide the constant the equation should be considered *improper.*

First find the H.C.F. of the dividend and the divisor which will be 1. For if it is not 1, it will divide the constant and after division, the dividend, divisor and the constant will be in standard form.

Go on dividing this standard dividend by the standard divisor till we get the remainder 1. Then write the quotients one below the other followed by the constant and 0. This column is called the Downward Column or 'Creeper'. Multiply the last but one number by the number over its head and add to this product the lower number and write this final resulting number over the head of the last but one number. By striking off the top and the end numbers, reduce the column. Repeat the procedure several times until we are left with two entries (at the top of the column). Then divide the upper number by the dividend. The remainder so obtained will be the quotient; this gives the value of y. Similarly the lower number divided by the divisor leaves a remainder which gives the smallest value of x.

In this procedure following changes may be necessary. If in the downward column, the number of divisions is even and the constant is positive, the remainders will be the values of x and y respectively. If the constant is negative and the number of divisions is even, then (Dividend – first quotient y) and (Divisor – first x) will be the desired y and x respectively.

(If the number of divisions is odd, just interchange.)

Comment: Now we'll solve the equations by Bhāskarācārya's method: $100x - 63y = -90$. Dividend = 100, Divisor = 63, H.C.F. = 1.

1	63	100	1
2	26	37	1
1	4	11	2
	1	3	

Divide bigger number by smaller, write the quotients in the right and left columns and remainders below.

This way we get six quotients – an even number: 1, 1, 2 in the right column and 1, 2, 1 in the left one. They are, in the order, 1, 1, 1, 2, 2, 1. Write these in a vertical column and write below the constant 90 and 0. $90 \times 1 + 0$ is written in front of 1. $90 \times 2 + 90 = 270$ is written above 90.

1	2430
1	1530
1	900
2	630
2	270
1	90
90	×
0	×

Continuing this way we write 630, 900, 1530 ending with 2430 at the top-most line. When 2430 is divided by 100, the remainder is 30 = y. When 1530 is divided by 63, the remainder is 18 = x. Now 100 × 18 – 63 × 30 = 1800 – 1890 = –90. So (18, 30) is a solution.

Complete solution:

x = 18	81	144	207	...
y = 30	130	230	330	...

If the equation is 100x – 63y = 90, we take
y = 100 – 30 = 70, x = 63 – 18 = 45.
Complete solution: (45, 70), (108, 170), (171, 270) . . . .

Next we'll see how to solve an equation when the number of divisions is odd.

यद्गुणा क्षयगषष्टिरन्विता वर्जिता च यदि वा त्रिभिस्ततः ।
स्यात्त्रयोदशहृता निरग्रकास्तं गुणं गणक मे पृथग्वद ।। CCLVII ।।

Solve: $\frac{-60x \pm 3}{13} = y$ in integers.

Comment: We'll first solve $\frac{60x + 3}{13} = y$. The table with 60, 13 is as on the left side. The sequence of quotients is 4, 1, 1, 1, 1 odd in number. Here the constant is 3. We write 4, 1, 1, 1, 1, 3, 0 in the left column and proceed from the bottom upwards as in the previous problem.

1	13	60	4
1	5	8	1
	2	3	1
		1	

4	69
1	15
1	9
1	6
1	3
3	×
0	×

Here 69 divided by 60 leaves the remainder 9 and y = 60 – 9 = 51. 15 divided by 13 leaves the remainder 2 and so x = 13 – 2 = 11. If the equation is $\frac{60x - 3}{13} = y$ we get x = 2, y = 9.

Example

एकविंशतियुतं शतद्वयं यद्गुणं गणक पंचषष्टियुक्।
पंचवर्जितशतद्वयोध्दृतं शुद्धिमेति गुणकं वदाशु तत्।। CCLVIII ।।

O arithmetician! 221 is multiplied by an integer; 65 is added to the product; 195 divides the result exactly. Tell the multiplier quickly. [Solve $\frac{221x+65}{195} = y$ in integers.]

Comment: Here we can divide by H.C.F. (221, 195) = 13 and get $\frac{17x+5}{15} = y$. Here the quotients for 17, 15 are 1, 7. The top two numbers are 40, 35. 40 divided by 17 leaves remainder 6 and 35 ÷ 15 leaves remainder 5. So x = 5, y = 6. Complete solution: (5, 6), (20, 23), (35, 40), (50, 57), ...

1	40
7	35
5	×
0	×

Example

भवति कुट्टकविधेर्युतिभाज्ययोः समपवर्तितयोरथवा गुणः।
भवति यो युतिभाजकयोः पुनः स च भवेदपवर्तनसंगुणः।। CCLIX ।।

If in $\frac{ax+b}{c} = y$, a, b have a common factor m, then

$$\frac{\frac{a}{m}x + \frac{b}{m}}{c} = \frac{y}{m} = y'$$

and solve this equation. Clearly values of x will be the same and y = my'.

Comment: We'll take $\frac{100x+90}{63} = y$ (stanza CCLI).

Divide by 10 to get

$$\frac{10x+9}{63} = \frac{y}{10} = y'.$$

Sequence of quotients for 10, 63 is 0, 6, 3 (odd).

0	27
6	171
3	27
9	×
0	×

27 ÷ 10 leaves the remainder 7;
171 ÷ 63 leaves the remainder 45.
Since the number of quotients is odd,
$x = 63 - 45 = 18$, $y = 100 - 10 \times 7 = 30$.

क्षेपजे तक्षणाच्छुद्धे गुणाप्ती स्तो वियोगजे।। CCLX ।।

The above methods are applicable when the constant is positive. If it is negative, take suitable complements (in a, b).

Example

यद्गुणा गणक षष्टिरन्विता वर्जिता च दशभिः षडुत्तरैः।
स्यात् त्रयोदशहृता निरग्रका तद्गुणं कथय मे पृथक् पृथक्।। CCLXI ।।

O arithmetician! an integer is multiplied by 60; 16 is subtracted from or added to the product. The results divided by 16 become residueless in each case. Tell me (those numbers) separately. [Solve: $\frac{60x \mp 16}{13} = y$ in integers.]

Comment: Here choosing +16 we get $\frac{60x + 16}{13} = y$. Since 16 > 13 we consider $\frac{60x + 3}{13} = y - 1 = y'$. As per (CCLVII), $x = 11$, $y' = 51$.

$\therefore$ $x = 11$, $y = 52$. To solve $\frac{60x - 16}{13} = y$, consider $\frac{60x - 3}{13} = y + 1 = y'$. Here $x = 2$, $y' = 9$,
$\therefore$ the solution is (2, 8).

Example

येन संगुणिताः पंच त्रयोविंशतिसंयुताः।
वर्जिता वा त्रिभिर्भक्ता निरग्राः स्युः स को गुणः।। CCLXII ।।

Five is multiplied by an integer; 23 is added to or subtracted from the product; 3 divides the results. Tell the multiplier. [Solve: $\frac{5x \pm 23}{3} = y$.]

Comment: We solve $\frac{5x \pm 2}{3} = y \mp 7 = y'$.

A solution of $\frac{5x + 2}{3} = y'$ is x = 2, y' = 4 and of $\frac{5x - 2}{3} = y'$ is x = 1, y' = 1.

So a solution of $\frac{5x + 23}{3} = y$ is x = 2, y = 11

and of $\frac{5x - 23}{3} = y$ is x = 7, y = 4.

Formula to solve the equation when the constant is zero

क्षेपाभावे तथा यत्र क्षेपः शुद्ध्येत् हरोद्धृतः।
ज्ञेयः शून्यं गुणस्तत्र क्षेपो हारहृतः फलम्।। CCLXIII ।।

If the constant (c) is zero or divisible by the divisor (b), a multiplier is zero and the (corresponding) quotient is the constant divided by the divisor (that is one solution is x = 0 and $y = \frac{c}{b}$).

Example

येन पंचगुणिताः खसंयुताः पंचषष्टिसहिताश्च तेऽथवा।
स्युस्त्रयोदशहृता निरग्रकाः तं गुणं गणक कीर्तयाशु मे।। CCLXIV ।।

The product of an integer and five is added to 0 or 65. The outcomes divided by 13 leave no residue. O arithmetician! tell me that number immediately. [Solve: (i) $\frac{5x}{13} = y$ and (ii) $\frac{5x + 65}{13} = y$.]

Comment: In (i), (0, 0) is one solution. Others are (13, 5), (26, 10), ...

In (ii), one solution is (0, 5). Others are (13, 10), (26, 15), ...

Formula: constant is 1

क्षेपं विशुद्धिं परिकल्प्य रूपं पृथक् पृथग्ये गुणकारलब्धी।
अभीप्सितक्षेपविशुद्धनिघ्ने स्वहारतष्टे भवतस्तयोस्ते।। CCLXV ।।

First solve $\frac{ax \pm 1}{b} = y$ and get a solution (x_0, y_0). Then a solution of $\frac{ax \pm c}{b} = y$ is: x is the remainder in $\frac{cx_0}{b}$ and y is the remainder in $\frac{cy_0}{a}$.

Comment: Suppose we wish to solve $\frac{17x+5}{15} = y$.

First solve $\frac{17x+1}{15} = y$ and get $x_0 = 7$, $y_0 = 8$.
$5 \times 7 \div 15$ leaves the remainder $5 = x$ and $8 \times 5 \div 17$ leaves the remainder $6 = y$.

Continued fractions provide a proof of Bhāskarācārya's method. To solve $ax - by = 1$ we form continued fraction for $\frac{a}{b}$,

e.g. $\frac{5}{13} = \frac{1}{13/5} = \frac{1}{2+\frac{3}{5}} = \frac{1}{2+}\frac{1}{1+\frac{2}{3}} = \frac{1}{2+}\frac{1}{1+}\frac{1}{1+}\frac{1}{2}$

If we omit the last one, viz., $\frac{1}{2}$ and compute the remaining fraction we get $\frac{1}{2+\frac{1}{1+1}} = \frac{2}{5}$ and we note that $13 \times 2 - 5 \times 5 = 1$ and thus $x = 2$, $y = 5$ is a solution of $13x - 5y = 1$. Similarly the penultimate convergent of $\frac{a}{b}$ provides a solution to $ax - by = 1$. Details can be found in an algebra text.

Formula: union of pulverizers with the same divisor

एको हरश्चेद्गुणकौ विभिन्नौ तदा गुणैक्यं परिकल्प्य भाज्यम्।
अग्रैक्यमग्रं क्रम उक्तवद्यः संश्लिष्टसंज्ञः स्फुटकुट्टकोऽसौ।। CCLXVI ।।

If an (unknown) integer is multiplied by two integers (separately) and the products divided by a (given) integer leave two remainders then (to find the unknown) assume the sum of the multipliers as dividend and the sum of the remainders as negative constant of a proper pulverizer, which is the union of two pulverizers.

Comment: If $ax/c = y + b/c$, i.e. $ax - b = cy$ ... (i)

and $a'x/c = y' + b'/c$, i.e. $a'x - b' = cy'$ ... (ii)

then $(a + a')x - (b + b') = c(y + y')$

is the union of two pulverizers (i) and (ii).

Thus if $ax \equiv b \pmod c$ and $a'x \equiv b' \pmod c$

then $(a + a')x \equiv b \pmod c$.

This is a well-known formula in the theory of congruences.

Example

कः पंचनिघ्नो विहृतस्त्रिषष्टया सप्तावशेषोऽथ स एव राशिः।
दशाहतः स्याद्विहृतस्त्रिषष्टया चतुर्दशाग्रो वद राशिमेनम्।। CCLXVII ।।

If the product of an (unknown) integer and 5 is divided by 63 then the remainder is 7. If the same integer is multiplied by 10 and divided by 63 then the remainder is 14. Tell that integer.

Comment: The united pulverizer is $(5 + 10)x - (7 + 14) = 63y$, i.e. $5x - 7 = 21y$. One solution is $(14, 3)$, and $x = 14, 35, 56, \ldots$

CHAPTER 34

Concatenation (Permutations, Partitions etc.)

स्थानांतमेकादिचयांकघातः संख्याविभेदा नियतैस्युरंकैः।
भक्तोऽङ्कमित्यांकसमासनिघ्नः स्थानेषु युक्तो मितिसंयुतिः स्यात्।। CCLXVIII।।

To find the number of permutations of given (n) different digits (or objects), write 1 in the first place, 2, 3, 4, . . . up to the number of objects (n) and multiply them. Divide the product of the number of permutations and the sum of the given (n) digits by the number of the given digits (i.e. by n); write the quotient the given number of times (i.e. n times) in a column but leaving one-digit place each time; add them; the result is the sum of the numbers formed (by permuting the given n digits).

Comment: The first part describes the familiar formula $_nP_n = n!. = 1 \times 2 \times 3 \times ... \times n$. It is known that $_nP_r = \frac{n!}{(n-r)!}$ is the number of permutations from n different objects taken r at a time. This is proved in algebra texts. It is easy to derive the formula for the sum of the numbers formed:

it is $\frac{(n-1)!\,(10^n - 1)\,(\text{sum of the digits})}{9}$.

Example

द्विकाष्टकाभ्यां त्रिनवाष्टकैर्वा निरंतरं ह्व्यादिनवावसानैः।
संख्याविभेदाः कति संभवन्ति तत्संख्यकैक्यानि पृथग्वदाशु।। CCLXIX।।

Using (i) 2, 8, (ii) 3, 8, 9, (iii) 2, 3, . . ., 9 how many different numbers can be formed? What is the sum of numbers, so formed, in each case?

Comment: (i) Only two numbers 28 and 82.

(ii) In this case $_3P_3 = 3 \times 2 \times 1 = 6$ different numbers.

(iii) Here the answer is $_8P_8 = 8! = 40320$.

In, (i) n = 2, sum of the digits = 10 and so the sum of the numbers formed = $\frac{99}{9} \times 10 = 110$.

(ii) n = 3, sum of the digits = 20 and so answer = $\frac{2 \times 999 \times 20}{9} = 4440$.

(iii) n = 8, sum of the digits = 2 + 3 + . . . + 9 = 44.

Answer = $\frac{7!\,(10^8 - 1)\,(44)}{9} = 2463999975360$.

Example

पाशांकुशाहिडमरूककपालशूलैः खट्वांगशक्तिशरचापयुतैर्भवन्ति।
अन्योन्यहस्तकलितैः कति मूर्तिभेदाः शंभोर्हरेरिव गदारिसरोजशंखैः।। CCLXX।।

Lord Śiva holds ten different weapons, namely a trap, a goad, a snake, a drum, a potsherd, a club, a spear, a missile, an arrow and a bow in his hands. [Lord Śiva has five heads and so it is presumed that he has ten arms.] Find the number of different Śiva idols. Similarly, solve the problem for Viṣṇu idols; Viṣṇu has four objects: a mace, a disc, a lotus and a conch.

Comment: In the first example, there are 10! = 3628800 plossible Śiva idols. In the case of Lord Viṣṇu, possible idols are 4! = 24 in number.
[*NOTE:* In the daily ritual (*sāndhyavandanam*) there are 24 names of Lord Viṣṇu.]

Formula: Permutations with repetitions

यावत्स्थानेषु तुल्यांङ्कास्तद्भेदैस्तु पृथक्कृतैः।
प्राग्भेदा विहृता भेदास्तत्संख्यैक्यं च पूर्ववत्।। CCLXXI।।

(To find the total number of permutations of given n digits or objects) if certain digits are alike then form the product of the number of permutations of those places at which alike digits occur (assuming that each block of alike digits has different digits), and divide the number of permutations of all the given (n) digits (assuming them different) obtained by the previous method (cf. stanza CCLVIII) by this product. And sum of the numbers formed is obtained by the previous method.
Comment: Let $a_1 = a_2 = \ldots = a_r$, $b_1 = b_2 = \ldots = b_s$, $c_1 = c_2 \ldots = c_t, \ldots$ be n objects. Then number of permutations $= \dfrac{n!}{r!\,s!\,t!\ldots}$, $n = r + s + t + \ldots$

This is proved in standard texts.

Example

द्विद्व्येकभूपरिमितैः कतिसंख्यकाः स्युः
तासां युतिं च गणकाऽऽशु मम प्रचक्ष्व।
अंभोधिकुंभिशरभूतशरैस्तथाङ्कैः
चेदंकपाशमितियुक्तिविशारदोऽसि।। CCLXXII।।

Find quickly the number of different numbers that can be formed with (i) 2, 2, 1, 1, (ii) 4, 8, 5, 5, 5. Also find their sums.

Comment: (i) $\frac{4!}{2!\,2!} = \frac{24}{4} = 6$.

In the unit's place 1 appears three times and so does 2. Their sum = 9. The same is true of digits in the ten's place, etc. and so the sum of the number = 9999.

(ii) Number of permutations = $\frac{5!}{3!} = 20$.

General formula for the sum

$$= \frac{n!}{r!\,s!\,t!\,n} \frac{(ra_1 + sb_1 + tc_1 + \ldots)}{9} (10^n - 1).$$

In (ii), n = 5, r = 1, s = 1, t = 3.

Sum = $\frac{5!}{3!\,5}$ $(4 + 8 + 15)(10^5 - 1) = 1199988$.

Example

स्थानषट्कस्थितैरङ्कैरन्योन्यं खेन वर्जितैः।
कति संख्याविभेदाः स्युर्यदि वेत्सि निगद्यताम्।। CCLXXIII।।

Leaving aside 0, if the digits 1 to 9 are written six at a time, then find how many different numbers are formed.

Comment: ${}_nP_r = {}_9P_6 = 9 \times 8 \times 7 \times 6 \times 5 \times 4 = 60480$.

निरेकमंकैक्यमिदं निरेकस्थानांतमेकापचितं विभक्तम्।
रूपादिभिस्तन्निहतेः समा स्युः संख्याविभेदा नियतेऽङ्कयोगे।। CCLXXIV।।

नवान्वितस्थानकसंख्यकायाः ऊनेऽङ्कयोगे कथिते तु वेद्यम्।
संक्षिप्तमुक्तं पृथुताभयेन नान्तोऽस्ति यस्माद् गणितार्णवस्य।। CCLXXV।।

When the sum (n) of the digits and the number (r) of blank spaces are given, follow the method given below to get the total number of permutations:

$\frac{(n-1)\ldots(n-r+1)}{(r-1)!}$. This method is applicable when n < r + 9. All this has been given briefly because this ocean of mathematics is vast.

Comment: This is an example of partitions and the derivation will be found in books on theory of numbers.

Formula can be written as $\frac{(n-1)!}{(n-r)!(r-1)!}$.

NOTE: The great Indian mathematician (Late) S. Ramanujan made outstanding discoveries in the theory of partitions. It is said that he got inspiration from the above stanza.

Example

पंचस्थानस्थितैरंकैर्यद्यद्योगस्त्रयोदश।
कतिभेदा भवेत्संख्या यदि वेत्सि निगद्यताम्।। CCLXXVI।।

Find the total number of different numbers of five digits whose sum is 13.

Comment: n = 13, r = 5 so the answer

$$=\frac{12!}{8!4!}=495.$$

न गुणो न हरो न कृतिर्न घनः पृष्टस्तथापि दुष्टानाम्।
गर्वितगणकवटूनां स्यात्पातोऽवश्यमंकपाशेऽस्मिन्।। CCLXXVII।।

Although there is no multiplication, division, squaring or cubing in this concatenation, yet if it is asked to egotistical evil-minded lads of astronomers, their humbling is certain.

येषां सुजातिगुणवर्गभूषितांगी
शुद्धाखिलव्यवहृतिः खलु कंठसक्ता।
लीलावतीह सरसोक्तिमुदाहरन्ती
तेषां सदैव सुखसंपदुपैति वृद्धिम्।। CCLXXVIII।।

This *Līlāvatī* clearly explains fractions, simple fractions, multiplication etc. It also beautifully describes problems in day-to-day transactions. Rules are transparent and examples are beautifully worded. Those who master this *Līlāvatī* will be happy and prosperous.

Alternate meaning: (Lass) Līlāvatī is born in a respectable family, stands out in any group of enlightened persons, has mastered idioms and proverbs. Whomsoever she embraces (marries) will be happy and prosperous.

अथ क्षेपकम् (INTERPOLATED VERSE)

अष्टौ व्याकरणानि षट् च भिषजां व्याचष्ट ताः संहिताः
षट् तर्कान् गणितानि पंच चतुरो वेदानधीते स्म यः।
रत्नानां त्रितयं द्वयं च बुबुधे मीमांसयोरन्तरे
सद्ब्रह्मैकमगाधबोधमहिमा सोऽस्याः कविर्भास्करः ।। CCLXXIX।।

Bhāskarācārya, the great poet and author of this book, had mastered eight volumes on Grammar, six on Medicine, six on Logic, five on Mathematics, four *Vedas,* a triad of three *ratnas,* and two (fore- and past-) *Mīmāṁsās.* He understood that the Lord (Superme) cannot be fathomed. [Obviously he had not mastered only Mathematics but was well versed in many branches of knowledge.]

Index of Verses

BEGINNING PART OF VERSE — STANZA NO.

Subject Index

[Brackets containing number(s) stand for the verse number(s)]